NATIVE AMF

NATIVE AMERICAN DNA

Tribal Belonging and the False Promise of Genetic Science

Kim TallBear

UNIVERSITY OF MINNESOTA PRESS
MINNEAPOLIS | LONDON

A different version of chapter 2 was published as "Native-American-DNA.com: In Search of Native American Race and Tribe," in *Revisiting Race in a Genomic Age*, ed. Barbara Koenig, Sandra Soo-Jin Lee, and Sarah S. Richardson (Piscataway, N.J.: Rutgers University Press, 2008), 235–52. Portions of chapter 4 were previously published as "Narratives of Race and Indigeneity in the Genographic Project," *Journal of Law, Medicine, and Ethics* 35, no. 3 (Fall 2007): 412–24.

Published by the University of Minnesota Press
111 Third Avenue South, Suite 290
Minneapolis, MN 55401-2520
http://www.upress.umn.edu

Library of Congress Cataloging-in-Publication Data

TallBear, Kimberly.
Native American DNA : tribal belonging and the false promise of genetic science / Kim TallBear.
Includes bibliographical references and index.
ISBN 978-0-8166-6585-3 (hc : alk. paper)
ISBN 978-0-8166-6586-0 (pb : alk. paper)
1. Indians of North America—Anthropometry. 2. Human population genetics—North America. 3. DNA fingerprinting—North America. 4. Genetic genealogy—North America. I. Title.
E98.A55T35 2013
970.004′97—dc23

2013012526

Printed in the United States of America on acid-free paper

20 19 18 17 16 10 9 8 7 6 5 4 3

Dedicated to my late grandmother,
Arlene Heminger-Lamb, a Dakota woman,
who symbolizes for me all those who think deeply on things,
but whose choices are few.

CONTENTS

ACKNOWLEDGMENTS

Many people helped along the path to this book. I can name but a few important individuals here, and I name them chronologically. I thank my mother, LeeAnn TallBear, for impressing on me from my earliest memory that education could make all the difference in living a full and productive life—that it could take me to interesting places in the world and that it could help me give back in the places where I have been rooted. As a single mother, she never said she could do it all, and she taught me early the practice of making kin. I also thank the late Vine Deloria Jr. I never met him, yet he started me thinking about the politics of anthropology in Native America before I could read. I asked my undergraduate student mother in 1973, "What does it mean, 'Custer died for your sins'?"

Thanks to community-planning gurus Marie Kennedy, Mel King, and Louise Dunlap for their investment in my planning education at the University of Massachusetts, Boston, and at MIT. I am grateful for their example that there is no contradiction in entangling knowledge building and community development.

Thanks to Mervyn Tano for employing me and introducing me to the Human Genome Project, where I began thinking about the ethical, legal, and social implications of genomics for indigenous peoples. I was hooked from the start.

Thanks to David S. Edmunds, a keen editor, a generative collaborator, my coparent and dear friend, for intellectual, moral, and financial support. He gave me countless hours to fill his ears with impassioned talk about genetics and race when he had his own work to do (or would rather have checked Michigan State basketball scores online). Thanks to Carmen TallBear-Edmunds for being the best baby girl in the world for us and for enduring all of our "politics talk." Thanks to Robin Johnson Weber and to Noriko (Ishiyama) and Jun Kamata for participating in nonbiological kin making and for nourishing that politics talk and the bonds entangling our families with food, conversation, laughter, and Zin.

Thanks to my advisers, James Clifford, Donna Haraway, Charis Thompson, and (unofficially) Sandra Harding, for theoretical tutelage, engaged critiques, and always treating me like a colleague. Thanks to Dakota language instructor Glen Wasicuna for encouraging my curiosity about the ontology embedded in the vocabulary and structure of the language. A second thanks to Jim Clifford for helping me articulate the value that I see in circulation, in "routedness" versus "rootedness," and to Donna Haraway for insisting that there is pleasure to be had in the confusion of boundaries—in their undoing. Jim's and Donna's works not only build me intellectually but also help guide me in living a life that I never could have imagined.

Thanks to legal scholar Rebecca Tsosie for her mentorship, her collaboration, and her example of dedication to Native American sovereignty, to supporting other indigenous scholars, and to rigorous intellectual analysis.

Thanks to Deborah Weiss Bolnick for enlightening conversation about the science and politics of ancient DNA and to Jonathan Marks for his skill in telling the big stories about the history of genetics and race. Thanks to Jenny Reardon for collaboration and for friendship. Like science fiction writers, Jenny and I write of a "shared world" and characters, sometimes entangling our narratives. This has led to fruitful collaboration. Thanks to my other coauthors (a de facto Listserv) on "The Science and Business of Genetic Ancestry Testing": Duana Fullwiley, Troy Duster, Richard Cooper, Joan Fujimura, Jonathan Kahn, Jay Kaufman, Ann Morning, Alondra Nelson, Pilar Ossorio, and Susan Reverby, for challenging conversations and for keeping us all up to date on literature and scandal. Thanks to Kalindi Vora, Neda Atanasoski, and Elly Teman for their editing earlier in the writing process. Thanks to my sister Jody TallBear for much-needed legal-research assistance related to the Kennewick Man narrative. Thanks to Carlos Andrés Barragán for translation of Genographic Project documents in Peru and for regional insight.

Thanks to the College of Natural Resources and the Department of Environmental Science, Policy, and Management at the University of California, Berkeley, for intellectual and financial support. Thanks especially to my wonderful colleagues in the Division of Society and Environment. Thanks to UC Berkeley graduate students with whom I have learned these past few years; work with them inspired me to tie methodological and theoretical threads together and get this book out the door: Hekia Bodwitch, Liz Carlisle, Clint Carroll, Rachel Ceasar, Erin Condit-Bergren, Robert Connell, Shannon Cram, Lindsey Dillon, Sibyl Diver, Ugo Edu, Zoe Friedman-Cohen, Sharon Fuller, Eric George (a brilliant undergraduate in my graduate seminar), Ted Grudin, Ellen Kersten, Esther Kim, Rosa-Maria Martinez, Paul Rogé, Ryan L. Shelby, Carolyn Smith, Beth Stephens (UC Santa Cruz and UC Davis), Chuck Striplen, Emma Tome (another brilliant undergraduate), Ali Tonak, Meredith Van Natta, Tammie Visintainer, Jesse Williamson, Cleo Woelfle-Erskine, and Kevin Woods.

Thanks to my fellow Native American and Indigenous Studies Association (NAISA) council members: Chris Andersen, Vince Diaz, Brendan Hokowhitu, J. Kēhaulani Kauanui, Jean O'Brien, Kate Shanley, Alice Te Punga Somerville, Maggie Walter, and Robert Warrior. They are models for me in doing theoretically incisive work that is always committed to indigenous sovereignty.

Thanks to my editors, Richard Morrison and Jason Weidemann, and to the rest of the skilled staff at the University of Minnesota Press for seeking me out and for seeing this book through to publication. Their faith in this project has been constant. Thanks to Stephanie Malia Fullerton, Jeffrey C. Long, Alondra Nelson, and one anonymous reviewer for helpful feedback on the manuscript. Thanks to Blaine Bettinger for feedback on the chapters on DNA-testing companies and genetic genealogy. Thanks also to Celeste Newbrough for her proficient indexing services on this book. I could not have done without them.

Thanks to the Sisseton-Wahpeton Oyate; the University of California, Berkeley; and the National Science Foundation for financial support. The views expressed here are not necessarily shared by my funders and advisers.

Introduction

AN INDIGENOUS, FEMINIST APPROACH TO DNA POLITICS

Scientists and the public alike are on the hunt for "Native American DNA."[1] Hi-tech genomics labs at universities around the world search for answers to questions about human origins and ancient global migrations. In the glossy world of made-for-television science, celebrity geneticist Spencer Wells travels in jet planes and Land Rovers to far-flung deserts and ice fields. Clad in North Face® gear, he goes in search of indigenous DNA that will provide a clearer window into our collective human past.

Others—housewives, retirees, professionals in their spare time—search for faded faces and long-ago names, proof that their grandmothers' stories are true, that there are Indians obscured in the dense foliage of the family tree. Some are meticulous researchers, genealogists who want to fill in the blanks in their ancestral histories. They combine DNA testing with online networking to find their "DNA cousins." Some have romantic visions of documenting that "spiritual connection" they've always felt to Native Americans. A few imagine casino payouts or free housing, education, and health care if they can get enrolled in a Native American tribe. Applicants to Ivy League and other top-ranked schools have had their genomes surveyed for Native American DNA and other non-European ancestries with the hope of gaining racial favor in competitive admissions processes. Former citizens of Native American tribes ejected for reasons having to do with the financial stakes of membership have sought proof of Native American DNA to help them get back onto tribal rolls.[2] One mother—herself an adoptee from a Native American biological mother—sought a DNA test in order to forestall legal termination of her parental rights. If she could represent herself and her child as genetically Native American, she hoped to invoke the Indian Child Welfare Act, which inhibits the adoption of children away from Native American parents and communities.[3]

What Is Native American DNA?

To understand Native American DNA, it is not enough to discuss simply what genetic scientists say they are looking for in their samples—though I will do that shortly. It is also important to look back at how Native American bodies have been treated historically, for knowledge-producing cultures and practices that shaped earlier research continue to influence the way science is done today. Biophysical scientists have for several centuries crafted and refined particular questions, terminologies, and methods in their studies of Native American and other marginalized bodies. Native American bodies, both dead and living, have been sources of bone, and more recently of blood, spit, and hair, used to constitute knowledge of human biological and cultural history. In the nineteenth and early twentieth centuries, the American School of Anthropology rose to worldwide prominence through the physical inspection of Native American bones and skulls plucked from battlefields or from recent gravesites by grave robbers–cum–contract workers for scientists. It was certainly distasteful work to scavenge decomposing bodies and boil them down so bones could be sent more easily to laboratories clean and ready for examination.

But two justifications emerged for the work, justifications that will ring familiar in my analysis of genetic scientists' treatment of Native Americans' DNA. First, this sort of research was and is for the good of knowledge, and knowledge, it was and is supposed, is for the good of all, despite complaints by Native Americans then and now about research purposes and methods. Second, the Indians were seen as doomed to vanish before the steam engine of westward expansion. Today, "indigenous peoples" are doomed to vanish through genetic admixture. The idea was then and is now that they should be studied before their kind is no more. It is not the means but the ends that science keeps its sights on.

Given that background, what, in technical terms, is Native American DNA? In the early 1960s, new biochemical techniques began to be applied to traditional anthropological questions, including the study of ancient human migrations and the biological and cultural relationships between populations. The new subfield of molecular anthropology was born, sometimes also called anthropological genetics.[4] Sets of markers or nucleotides in both the mitochondrial DNA (mtDNA) and in chromosomal DNA were observed to appear at different frequencies among different populations. The highest frequencies of so-called Native American markers are observed by scientists in "unadmixed" native populations in North and South America. These markers are the genetic inheritance of "founder populations," allegedly the first humans to walk in these lands that we now call the Americas.

On the order of millennia, anthropological geneticists want to understand which human groups, or "populations," are related to which others, and who descended from whom. Where geographically did the ancestors of different human groups migrate from? What were their patterns of geographic migration, and when did such migrations occur? In the genomes of the living and the dead, scientists look for molecular sequences—the "genetic signatures" of ancient peoples whom they perceive as original continental populations: for example, Indo-Europeans, Africans, Asians, and Native Americans. Native American DNA, as a (threatened and vanishing) scientific object of study, can help answer what are, for these scientists, pressing human questions.

Unlike scientists or consumers of genetic-ancestry tests, I stretch the definition of Native American DNA beyond its usual reference to "New World" genetic ancestry traceable either through female mtDNA and male Y-chromosome lines or through more complex tests that combine multiple markers across the genome to trace ancestry. I include the "DNA profile" as I examine the material and social work that Native American DNA does in the world. Commonly used in criminal cases, this test has been referred to as a "DNA fingerprint." Within an individual's genome, multiple sets of genetic markers are examined. They act like a genetic fingerprint to identify an individual at a very high probability. As a parentage test, the same form of analysis shows genetic relatedness between parent and child. When used by U.S. tribes and Canadian First Nations as part of conferring citizenship (also called "enrollment"), the DNA fingerprint becomes essentially a marker of Native Americanness.

More than genetic-ancestry tests that target "Native American" as a race or panethnic category, the DNA profile is helping to reconfigure the concept of tribe.

Technically, the DNA profile promises only to identify an individual or close biological kin relationship. But one must have a basic grasp of several types of complex knowledges simultaneously—molecular knowledges and their social histories, and practices of tribal citizenship—or the DNA profile is likely to be taken as a powerful marker of Native American identity. Those who understand its technical limitations—say, DNA-testing-company scientists and marketers—do not have a deep historical or practical understanding of the intricacies of tribal enrollment. Nor do they tend to understand the broader political frame circumscribing their work, how their disciplines have historically fed from marginalized bodies. Tribal folks know these politics and histories well—we live day in and day out with enrollment rules, and we all know about the Native American Graves Protection and Repatriation Act (NAGPRA)—but we do not know the molecular intricacies of the test. Where knowledge is lacking, gene talk—the idea that essential truths about identity inhere in sequences of DNA—misleads us. DNA tests used by tribes are simply statements of *genetic* parentage that tribal governments have made regulatory decisions to privilege instead of or along with other forms of parent-child relationship documentation, such as birth or adoption certificates. Tribes increasingly combine DNA tests with longer-standing citizenship rules that focus largely on tracing one's genealogy to ancestors named on "base rolls" constructed in previous centuries. Until now, tribal enrollment rules have been articulated largely through the symbolic language of "blood." Like many other Americans, we are transitioning in Indian Country away from blood talk to speaking in terms of what "is coded in our DNA" or our "genetic memory." But we do it in a very particular social and historical context, one that entangles genetic information in a web of known family relations, reservation histories, and tribal and federal-government regulations.

A Culmination of History and Narrative

Clearly, mtDNA lineages A, B, C, and D, and X- or Y-chromosome lineages M, Q3, and M3 are not simply objective molecular objects. These molecular sequences, or "markers"—their patterns, mutations, deletions, and transcriptions that indicate genetic relationships and histories—have not been simply uncovered in human genomes; they have been conceived in ways shaped by key historical events and influential narratives. Native American DNA as it is usually defined refers to molecules that track deep genetic and geographic ancestries (sometimes they code for genetic traits, often not) amplified from blood or saliva—less often from bone and hair—via chemicals and laboratory devices. The concept of Native American DNA is also conditioned by complex software that calculates frequency distributions of markers among different populations of the world from whom biological samples have been taken.[5]

But Native American DNA could not have emerged as an object of scientific research and genealogical desire until individuals and groups emerged as "Native American" in the course of colonial history. Without "settlers," we could not have "Indians" or "Native Americans"—a pan-racial group defined strictly in opposition to the settlers who encountered them. Instead, we would have many thousands of smaller groups or peoples defined within and according to their own languages, as Diné, Anishinaabeg, or Oceti Sakowin, for example. It is the arrival of the settler in 1492 and many subsequent settlements that frame the search for Native American DNA before it is "too late," before the genetic signatures of the "founding populations" in the Americas are lost forever in a sea of genetic admixture.

Of course, mixing is predicated on the notion of purity. The historical constitution of continental spaces and concomitant grouping of humans into "races" is the macro frame of reference for the human-genome-diversity researcher. Scientists who trace human migrations do not tell a story from the standpoint of those peoples who were encountered; they tell a story from the standpoint of those who did the encountering—those who named and ordered many thousands of peoples into undifferentiated masses of "Native Americans," "Africans," "Asians," and "Indo-Europeans." Standing where they do—almost never identifying as indigenous people themselves—scientists who study Native American migrations turn and look back over their shoulders with desire to know the "origins" of those who were first encountered when European settlers landed on the shores of these American continents.

In human genome diversity research, faith in the origins gets operationalized as "molecular origins." This refers to ancestral populations that are inferred for an individual based on a specific set of genetic markers, a specific set of algorithms for assessing genetic similarity, and a specific set of reference populations.[6] But each of those constitutive elements operates within a loop of circular reasoning. Particular, and particularly pure, biogeographic origins must be assumed in order to constitute the data that supposedly reveals those same origins. Native American DNA as an object could not exist without, and yet functions as a scientific data point to support the idea of, once pure, original populations. Notions of ancestral populations, the ordering and calculating of genetic markers and their associations, and the representation of living groups of individuals as reference populations all require the assumption that there was a moment, a human body, a marker, a population back there in space and time that was a biogeographical pinpoint of originality. This faith in originality would seem to be at odds with the doctrine of evolution, of change over time, of becoming.

The populations and population-specified markers that are identified and studied mirror the cultural, racial, ethnic, national, and tribal understandings of the humans who study them. Native American, sub-Saharan African, European, and East Asian DNAs are constituted as scientific objects by laboratory methods and devices, and also by discourses or particular ideas and vocabularies of race, ethnicity, nation, family, and tribe. For and by whom are such categories defined? How have continental-level race categories come to matter? And why do they matter more than the "peoples" that condition indigenous narratives, knowledges, and claims?

The answer to this last question is not because favored scientific categories are more objectively true. Privileging the concept of genetic population enables the sampling of some bodies and not others. An Anishinaabeg with too many non-Anishinaabeg ancestors won't count as part of an Anishinaabeg "population." To make things even more complicated, a scientist may draw blood from enrolled members of the Turtle Mountain Band of Chippewa Indians at a reservation in North Dakota and call her sample a "Turtle Mountain Chippewa" sample. At the same time, she may have obtained "Sioux" samples from multiple other scientists and physicians who took them at multiple sites (on multiple reservations or in urban Indian Health Service hospitals) over many years. In the first instance, we have a "population" circumscribed by a federally recognized tribal boundary. In the second, we have a "population" circumscribed by a broader ethnic designation that spans multiple tribes. There is often little categorical consistency between different study samples. That is because samples are delineated and named differently depending on where they are obtained and on how that government or institution organizes its citizenry or service population. There are histories of politics that *inhere in* the samples. Added to that are the politics *imposed onto* the samples by researchers who enforce subsequent requirements for the data, namely, that usable samples come only from subjects who possess a certain number of grandparents from within said population.

But such problems have done little to undermine the authority of scientists on questions of Native American origins and identity that precede study of our genome diversity. In the "real world" of power and resource imbalances, in which some peoples' ideas and knowledge are made to matter more than others, genetic markers and populations named and ordered by scientists play key roles in the history that has come to matter for the nation and increasingly the world. If such narratives are rescripting what is historically salient, they risk rescripting what is socially and politically salient with real material consequences. Native American DNA is material-semiotic.[7] It is supported by and threads back into the social-historical fabric to (re)constitute the categories and narratives by which we order life. Indigenous political authorities and identities, as well as land and resource claims, are at stake.

Who Studies? Who Gets Studied?

This book draws from well-documented histories of the science of race in the West, tracing a genealogy of Native American DNA as a research object and tool for categorizing molecules and humans. The book then examines an online community of "genetic genealogists" who use DNA to help trace family histories, before crossing into a strange hybrid world where science meets corporate marketing. The chapters reveal a gold mine of representational language and imagery excavated from company Web sites, narratives of origins, race, and tribe as told to us by DNA-testing companies and the Genographic Project.

Native American "perspectives" on genetic research or understandings of DNA are not a chief topic of this book. There is a rich field of data to mine in the evolution in Indian Country of blood into gene talk during the last ten years. When I am back in South Dakota on one of the two Dakota reservations that I call home—those of the Sisseton-Wahpeton Oyate and the Flandreau Santee Sioux tribes—or when I travel throughout Indian Country on business or for pleasure, I attend gatherings large and small: tribal membership meetings in community centers or casino banquet halls, powwows, tribal and federal agency meetings, tribal enrollment conferences, national tribal or First Nations organization meetings, and my annual tribal writers' retreat in southeastern South Dakota. Or I might simply enjoy a coffee and some gossip with relatives or friends in a reservation border-town café. In doing so, I regularly encounter tribal folks old and young, university-educated and not, and of different class backgrounds who refer to certain characteristics being "in our blood" or increasingly "part of our DNA" or our "genetic memory." Genetic memory refers to a sense of ancestral memory. That is, one might know a place or have knowledge of a place and the nonhumans found there that was not gained actively or personally. Rather, one somehow carries or embodies such knowledge or has a sense of having been in a place before because of ancestors' historical experiences of that place. One often hears such accounts in relation to the idea that descendants continue to retain knowledge or a sense of deep familiarity with place in spite of their ancestors' dispossession from the land and from tribal languages. Or descendants might have inherited "in our DNA" historical trauma from ancestors that continues to hamper individuals in their daily lives decades later. This sense of inexplicable inheritance would not have been chalked up to genetics twenty years ago. It would have been spoken of in the language of blood. Later chapters of this book discuss blood as not simply a biophysical substance in contemporary Native American parlance. Likewise, the new "DNA talk" in Indian Country works much the same way. Like blood, DNA gets spoken of as a more-than-biological substance. Yet, with money and resources at stake, DNA is also increasingly spoken of in indigenous communities as a scientifically objective and precise solution to an intractable political problem: who gets to be a tribal or First Nation citizen? As Jessica Bardill puts it, "DNA concretizes that idea ["Indian by blood"] and removes its ability to be a metaphor . . . only making it possible to mean the literal substance."[8]

Without ethnography, what indigenous peoples' own transitions from blood to DNA talk mean for indigenous ontologies or for citizenship practices is difficult to pin down. These developments would make a fascinating study. In fact, that is part of the aim of Bardill, who is affiliated with the Eastern Band of Cherokee, in her look at that tribe's enrollment practices. But this book, in perhaps an extreme form, commits what Audra Simpson calls an "ethnographic refusal," that is, "a calculus ethnography of what you need to know and what I refuse to write in."[9] Put simply, Native Americans are less subjects of this book than a key part of its audience. In an *explicitly* ethical move and from an *explicitly* situated place (this author stands at times in a lot of different Native American communities), this work refuses ethnography on Native Americans and instead gazes upon those who are understudied yet influential, those who track Native American DNA in bodies across time and space: genetic scientists (including anthropological geneticists), commercial enterprises, and financially able lovers of genetic science.

Indeed, it is not primarily Native American perceptions or (mis)-understandings of biology, genetics, and human genome diversity that matter in the delineation of Native American identities and attendant rights to self-governance. Tribal sovereignty aside, dominant U.S. understandings of race, kinship, history, and Native American identity set the ground upon which tribal and First Nations attempt to govern their citizenries and territories. Understanding what Native American DNA portends for Native Americans requires, in large part, understanding how gene discourses and scientific practices are entangled in ongoing colonialisms. What "they" think and do have always determined how much trouble "we" have.

Native American intellectuals, policy makers, and concerned tribal or First Nation citizens should find much of interest in these pages. Genetic understandings of history and identity can operate without reference to the federal-tribal legal regime that is critical for contemporary indigenous governmental authority, including rights to determine citizenship. In some instances, DNA testing can support indigenous governance, such as in the case of using a parentage test as part of a suite of enrollment or citizenship criteria. But an increasing geneticization of the categories of tribe, First Nation, and race can also have undesired consequences. "Genetic-ancestry" tests are irrelevant to existing indigenous citizenship criteria, whereas across-the-membership application of "parentage" tests can contradict hard-won legal foundations that are the source of contemporary indigenous governmental authorities. In other cases, enrollment controversies result from contradictions between categories, for example, when notions of "tribal nation" and "race" do not neatly line up but collide. This was the matter in the Cherokee and Seminole freedmen cases, where African American tribal members, descended from freed slaves naturalized as tribal citizens long ago, had their citizenship questioned or revoked due to their nonblood ties to the tribes.[10] Bringing DNA to bear on such identity and citizenship claims will complicate and not solve disagreements rooted not only in biology but also in histories of white supremacy, slavery, and dispossession. The question is, as genetic identities and historical narratives command increasing attention in society, will they come to rival as legitimate grounds for identity claims the existing historical-legal foundations of indigenous governance authority? In the United States, that authority is treaties and case law. If so, we will see a transformation, not an end, to controversy in indigenous citizenship and Native American racial identity, adding to a growing genetic fetishism in the broader society.

How I Produce My Knowledge: Theory, Method, and Ethics

The knowledge foundations of this book are diverse and not contained by academic disciplines. But from within the academy, this book brings into conversation voices and theories from science and technology studies (STS), or social studies of science and technology, and Native American and indigenous studies (NAIS). It also draws on "cultural studies" scholarship and frameworks but informs them with STS and NAIS literatures and methods. Cultural studies was at the forefront of problematizing the "tradition"-versus-"modernity" binary that does so little to illuminate the questions tackled in this book. Its critical social theories and interpretive textual approaches assist a focus on *both* scientific and indigenous spaces as cultural, political, and knowledge-production spaces. In turn, this book relies on social-science methods such as participant observation and interviews to complement interpretive claims.

All three fields—STS, NAIS, and cultural studies—share critiques of universality and objectivity in the Western sciences, with feminist-oriented strands of STS being more critical in that regard than is mainstream STS.[11] STS uses social-science and humanities approaches to explain how social, political, and cultural values affect science-and-technology (technoscientific) research and innovation and, in turn, how technosciences affect our politics, cultures, and social institutions.

The following chapters focus on challenges to indigenous governance posed by human genome diversity research and its commercialization. But ultimately the goal of this book is to center the roles of the social and technosciences in *expanding* indigenous governance, in part through indigenous efforts to govern technoscientific knowledge production such that indigenous interests are protected. The exercise of indigenous sovereignty in the twenty-first century depends in no small part on how indigenous peoples account for the roles of technoscientific knowledge production. This book points in its concluding chapter to efforts by Native American tribes, Canadian Aboriginal peoples, and indigenous advocates to shift power relations in technoscientific knowledge production by asserting rights of property and control over the biological resources, processes, and material and conceptual objects of genome science.

Analytical Frameworks: Coproduction (of Natural and Social Orders) and Articulation

Readers will frequently encounter the idiom of "coproduction," a key STS analytical tool that explains natural and social orders as coproduced. That is, science and technology are explained as actively entangled with social norms and hierarchies. Rather than being discrete categories where one determines the other in a linear model of cause and effect, "science" and "society" are mutually constitutive—meaning one loops back in to reinforce, shape, or disrupt the actions of the other, although it should be understood that, because power is held unevenly, such multidirectional influences do not happen evenly.[12]

The emergence of new "natural orders" (for example, population categories such as "Native American" supported by genetic markers found in higher and lower frequencies in different geographies) depends in part on an already existing "social order" (for example, a concept of indigeneity and actual indigenous social organization) that itself was already informed over the centuries by natural orders (for example, race categories based on older notions of blood configured through material observation and symbolic meanings simultaneously). This twenty-first-century "natural order" (DNA markers labeled "Native American") continues the coproduction loop by helping to reconfigure as a genetic category the preexisting social and otherwise biological category of "indigeneity" that informed its emergence.

Coproduction facilitates and helps make sense of what could otherwise be a very confusing multidisciplinary analysis of the emergence of Native American DNA as a complex social and scientific object.[13] In terms of Native American DNA, it problematizes a realist approach to understanding the object. Native American DNA is not simply "naturally" determined; it becomes manifest as scientists observe the movement of particular nucleotides via human bodies across time and space (between what is today Siberia/Asia and the Americas). The presence of such markers is then used to animate particular "populations" and individuals and their tissues (both dead and living) as belonging to that identity category. The presence of such DNAs is sometimes also used by scientists and consumers to forge consequential identity connections between human bodies across time and space. But such bounded ethnic or racial descriptions of certain nucleotide sequences would not have any salience were it not for the established idea within genetic science that "Native American" (or "Amerindian" and the like) is a distinct genetic or biological category. And although researchers increasingly acknowledge indigenous governance authority (they are required to as they seek tribal-council or institutional-review-board approval to do research within Native American lands and on native bodies), at a more fundamental or conceptual level, the assumptions, technical languages, and methods of these fields *cannot* recognize indigenous political organization or identity. Indeed, they must actively work to eliminate social and biological complexity from their samples, as it is seen to interfere with the goal of tracing the migrations of nucleotides (that come to stand for the migrations of peoples) from supposedly "original" populations to effect "colonizations" of the "Americas." Thus, "Native American" becomes a moniker used to represent a clearly traceable biological link to the

"Old World" that lies back beyond the Bering Strait, rather than a label indicating long-standing and intimate relationships between humans and nonhumans on this side of the Bering Sea—relationships that cohere peoples as peoples with origins in specific landscapes.

The concept of articulation is also invoked in these pages, a framework traced to two prominent cultural studies scholars, James Clifford and, before him, Stuart Hall.[14] Like coproduction, the articulation framework complicates overly dichotomous views of phenomena as either essentially determined or overly constructed or invented, thereby implying a lack of "realness." In Clifford's and Hall's more specific terms, "articulation" (literally, to conjoin parts together into something neither strictly old and traditional nor completely new and different) indicates cultural transformation. It focuses on cultural practices and knowledges as sometimes borrowed, interpreted, and reconfigured. Indeed, dynamism in cultural practice and identity formation is a sign of being alive, another key claim that indigenous peoples consistently make. They have survived. They are still here.

A second key component of the articulation framework, and the reason it is helpful in relation to this subject, is its focus on power. It enables an analysis of dynamic societal forces that combine to determine who or what counts as "indigenous," including the power of genomic practices and articulations to structure indigenous lives, and in ways that may ultimately harm indigenous peoples more than they serve them. Perhaps the genetic articulation of indigeneity will become also an *indigenous* articulation as tribes increasingly move toward DNA parentage testing within a discourse of sovereignty.[15] Together, the frameworks of articulation and coproduction help us to see the emergent and not inevitable quality of genetic forms of indigeneity.

Central points of contention in this book reflect these two approaches. Those points of contention lie in the fact that particular questions posed in the course of human genome diversity research derive from particularly situated inquirers. And just as particular hypotheses are not equally relevant across diverse societal terrains, not all methods work equally well for differently positioned social actors. Let us turn now to a closer exploration of such ethical and methodological choices and the ways they shape the account in the chapters to follow.

The Sins and Stories of Anthropology: Embraces and Refusals

In *Custer Died for Your Sins* (1969), Vine Deloria Jr. famously interrogated the practices and power of anthropology to define and represent Native peoples' histories, practices, and identities to the world. Deloria was critical of various anthropological "revelations" of Indians as "folk people," as "caught between two worlds," or as not "real" enough because they didn't do enough Indian dancing. He saw such representations and the influence of anthropology on the American popular imagination, including in Indian Country, as detrimental to Indian self-concepts and notions of self-sufficiency and to Indian political organizing, assumptions of power, and self-governance.[16]

I grew up in the early 1970s in eastern South Dakota among Native American undergraduates, artists, and activists. Deloria's text and its reception in my corner of Indian Country shaped my relationship to anthropological representations and to the very idea of research from early on. It has become second nature for me to ponder the politics that run through knowledge production at every stage: how authors and researchers begin where they do, which audiences they imagine will receive their knowledge production, and what leads them to assume that they should research a subject or object. I wonder how researchers gain access to subjects, who brokers their research relationships, how much it costs to do research, and who funds research. I think about which research protections are in place and whose certifications, laws, and policies guide those protections. I wonder who controls and has access to data and whose languages are at play.

When I speak of anthropology, I speak of all of the subdisciplines: sociocultural, physical (or biological), archaeological, and linguistic anthropology. I have been most intimate with the first two subfields, and I draw most of my examples from those. James Clifford and George Marcus, in their seminal edited volume *Writing Culture* (1986), foreground ethnographic writing as cultural, as producing a cultural form. "Science is in, not above, historical and linguistic processes," they note.[17] It is itself cultural practice. Writing is central to what anthropologists do. Thus, their writing cannot be viewed simply as studied or distanced and unproblematic representations of the cultural practices and beliefs of the others they study, but must be read as a cultural-political act in and of itself, as a literary act. Ethnography does not render simply transparent representations of what is culturally "out there." For Clifford and Marcus, ethnography is "actively situated *between* powerful systems of meaning," posing its questions at the borders "of civilizations, cultures, classes, races, and genders." The same can be said of physical anthropology with its own specific methods, language, and texts that attempt to capture human movements and cultural histories via intimate examination of human bodies. Both forms of anthropology do not gaze from outside the cultural processes they represent; they are part of those very processes. Anthropology "decodes and recodes."[18] Ethnography and physical anthropology are translation. In that is power. What happens when a lifelong critic of anthropology is faced with the fact that one subdiscipline of anthropology is an incisive way to analyze—in the interest of Native American self-determination—the colonial practices and power of another anthropological subdiscipline?

I began with trepidation research on Native American DNA. The idea of constituting reservations where I have lived and worked, where I attend family and tribal events, where my family lines trace to, as "field sites" felt extractive. For this book, I conducted only a handful of interviews in Indian Country. Putting that informed-consent form between me and other tribal members felt wrong, like making an object of Indian Country rather than "routing" me in and through it.[19] I could not bring myself to write about my fellow tribal citizens, our family histories having been entwined for centuries, especially if that writing took place largely outside a shared or community-based work project. My queasiness with academic social-science research had to do with feelings of the individual power I could exercise via standard academic approaches to extract knowledge and to publish analyses under my individual name, to speak *for* versus speaking *with*.

My position on the continuum of indigeneity is not unimportant: I left the reservation at fourteen for high school in a progressive urban area and went to university on the East and West Coasts. As I inhabit diverse worlds, I confront the inequalities of race, class, and increasingly gender privilege from a very particular and perhaps unusual position. I am a privileged U.S. American—from the heart of empire—yet from one of the most impoverished groups here.

Moving in and out of multiple disciplinary, national, and ethnic cultural spaces, I work in an era of critical methodological interventions and hybrid writing styles that enable self-reflexivity and polyvocality in ethnography. For example, analyses of the subjectivity of the "endogenous," "native," "indigenous," or "insider" anthropologist, although they are radical in the history of anthropological inquiry, are limited.[20] Such analyses take up the benefits and risks of the position. On the one hand, an insider anthropologist might more quickly immerse herself in a community, gaining better access to insider data because of a more intimate familiarity with, as Emiko Ohnuki-Tierney puts it, the "cognitive and emotive dimensions" of subjects' behavior.[21] Sometimes research subjects are more willing to participate because they want to support their fellow—the insider anthropologist—and her career aspirations. On the other hand, a potential insider—one with a university education—can be suspect, classified as an "educated fool." If an insider anthropologist does gain a certain level of acceptance and support, she might be faced with yet another challenge, how to manage increased personal obligations and requests for reciprocity. Neither clearly insider nor outsider, the researcher experiences "gradations of endogeny."[22]

But such analyses do little to alleviate my unease. They are primarily concerned with matters internal to the discipline, an individual researcher's problematic subjectivity and her ability to produce rigorous disciplinary knowledge nonetheless. The discipline remains both the chief audience for such analyses and the chief agent in knowledge production. The agency and desires of subjects and communities to produce knowledge they need and in ways amenable to their ethical sensibilities are not central to such discussions. In preparing this book, I ultimately refused the role of the native (on native) anthropologist.

Indigenous scholars Pakki Chipps and Audra Simpson call for "refusal" of anthropology for what we might call ethical reasons. In certain moments, when occupying both roles simultaneously and proficiently was not possible, they chose being in solidarity or in sync with the demands and practices of their native communities first. Chipps refused research when it disrupted her role in the family or tribe, for example, the cooking and visiting, the doing instead of asking questions. For Simpson, her refusal came in the depths of anthropological process, when anthropology risked subverting sovereignty by compromising the "representational territory gained" by Natives. Simpson writes that there are things she will not write, understandings between her and her "informants"—her fellow First Nation citizens—that are implicit rather than explicit. She writes of turning off the tape recorder, of deciding she's revealed enough and respecting that her informants have revealed enough.[23]

To use Simpson's term, my "calculus ethnography,"[24] or what I refused to write, became almost everything to do with Native American perspectives on blood and DNA and their roles in forming contemporary Native American citizenship and identity. The concept of refusal helps frame the silences in this book as not only against the ethnographic grain but as productive and supportive of indigenous self-determination. After all, on the one hand, who among my tribal kin asked me to produce knowledge about their understandings of DNA and how molecular knowledge contradicts or might be compatible with their views of life, identity, and history? No one. On the other hand, through years of work as a tribal and environmental planner I felt certain of an implied support for research I might do that could support tribal governance and community development in relation to DNA knowledge production. Could I do such research without making my tribal kin my subjects?

This book is a response to that question. It shifts the anthropological and analytical gaze to non-Native subjects and scientific projects. Not only was shifting the gaze personally liberating, a relatively unimportant development in the grand scheme of things, but it also facilitated what I hope the reader will agree is an incisive account of Native American DNA as a material-semiotic object with power to influence indigenous livelihoods and sovereignties, and genetic scientists and entrepreneurs as frontline agents in the constitution of that object. Indigenes themselves, it turns out, were never the "key informants." That is not to say that observations made in Indian Country are absent from this work. I could not keep this narrative clean of things I've learned in a lifetime of living and working in various tribal and urban Indian communities. It is that life which makes this topic so compelling to me in the first place. That is the standpoint from which I begin, but beginning there, I look "up" and out.

In refusing to occupy the role of the native anthropologist, I made a more typical anthropological decision to study those who are culturally foreign to me. Yet that move is simultaneously unexpected. I am a Native American studying non-Natives. And I study "up." In the same year that Vine Deloria Jr. published his groundbreaking essay "Anthropologists and Other Friends" (1969), Berkeley anthropologist Laura Nader published her now-classic essay "Up the Anthropologist," in which she urged anthropologists to study "the colonizers rather than the colonized, the culture of power rather than the culture of the powerless, the culture of affluence rather than the culture of poverty."[25] Nader's admonition is not inconsistent with science-and-technology studies approaches. Because relations of power script my (un)ease with research, I feel easier with the ethics of studying scientists than studying Native Americans. As geographer and political ecologist Paul Robbins writes, "Research is theft,"[26] and from natives much has already been taken.

I thought I would find it morally and personally easier to study *not* the marginalized native but actors at the center of genomic fields—subjects with institutional, cultural, and financial power that enables them to develop and use genomic technologies for their own intellectual and commercial projects. I would not have to study those who engage with genomics chiefly as its potential research subjects—those who extend arms to receive the needle while the syringe receives their blood. In practice, it has not been easier to study up. Power inequities are not always clear. Genetic scientists are increasingly women, Latino, African or African American, and even Native American. Many are young and working hard to do scientific work as people of color and/or women in a predominantly white, male world. When I gaze upon scientists as individuals, when I take up their knowledge production and turn it over in my hand—an artifact ripe for cultural analysis—I am not unconscious that the social sciences and humanities are, like the natural sciences and engineering, born of colonialisms. I, too, have some power and privilege in my authorial position. In general, the natural scientists whom I study and by whom I am surrounded enjoy more cultural power in part because they command greater wealth in the form of government, foundation, and corporate research money. A humanities and social-science scholar in a largely natural-science department, I benefit from the margins.

"Decolonizing" Research

Several more research approaches (at least, their underlying ethics if not actual methods) that can be grouped under the growing umbrella of "decolonizing methodologies," to borrow Linda Tuhiwai Smith's influential terminology,[27] inform the development of this account. The "participatory action research" (PAR) movement, also commonly called "community-based participatory research" (CBPR),[28] emerged in the 1970s in developing countries as a critical and productive response to dominant research paradigms that extracted knowledge for the benefit of researchers, institutions, and governments, often at social and material cost to the peoples whose bodies, resources, and societies were the objects of study.[29] Combining research, education, and action, participatory research quickly gained currency around the world and is now used across scientific fields. In U.S. tribal communities, participatory techniques are often used in health-related research.[30] It is meant to address uneven power relations in research, especially on indigenous peoples, poor people, and other marginal groups.

Louise Fortmann, a feminist political ecologist who has conducted community forestry research in West Africa, focuses in part on the gendered relationships between humans and forest resources. She sees the practice of writing under a singular byline, as, for example, "I would like to thank so-and-so for their patient and thorough research assistance," as giving insufficient credit. If, she challenges, "local knowledge really is important and not just something we pay lip service to, then we should pay for it in our own currency—not with an offhand acknowledgment in a footnote but with full-blown academic credit. If we have relied on their knowledge, then they should be co-authors."[31] Fortmann and the community researchers she worked with have done just that.[32]

But Fortmann does not just trade in the researcher's most valuable currency, shared authorship in academic publications; she also works to expand the definition of what counts as currency—as legitimate research and knowledge production. She works to make research a more valuable currency for the communities with whom she studies. Fortmann explains that people "develop a consciousness about their problems as they [talk] about them." They become experts and spokespeople, and they develop networks that can later be mobilized to local ends.[33] In human genome diversity research, shared authorship with Native American subjects may be a step forward,[34] but a truly mutually beneficial relationship entails reconceptualizing research projects, stretching them beyond academic or corporate researchers' knowledges and institutions to also develop communities' broader intellectual property and their educational, economic, and governance institutions.

Because of its commitments to capacity building and to producing knowledge that communities expressly desire, CBPR has become a mantra for indigenous researchers, and increasingly for researchers working in indigenous communities.[35] "Decolonizing methods" deepen further still CBPR's ethical commitments to communities. Rather than integrating community priorities with academic priorities, changing and expanding both in the process, decolonizing methods begin and end with the standpoint of indigenous lives, needs, and desires, engaging with academic lives, approaches, and priorities along the way. Smith's *Decolonizing Methodologies* (1999) opens with a classic charge against researchers by indigenous peoples. Because it captures so completely the critiques of this field, I quote Smith at length:

> From the vantage point of the colonized . . . the term "research" is inextricably linked to European imperialism and colonialism. The word itself, "research," is probably one of the dirtiest words in the indigenous world's vocabulary. . . . Just knowing that someone measured our "faculties" by filling the skulls of our ancestors with millet seeds and compared the amount of millet seed to the capacity for mental thought offends our sense of who and what we are. . . . It angers us when practices linked to the last century, and the centuries before that, are still employed to deny the validity of indigenous peoples' claim to existence, to land and territories, to the right of self-determination, to the survival of our languages and forms of cultural knowledge, to our natural resources and systems for living within our environments.[36]

Smith, a Maori writing in New Zealand, centers indigenous perspectives on research in order to deconstruct Western scientific research and to reiterate its role in the imperial project. But she does not stop there. Smith advocates a Kaupapa Maori methodology in which Maori "assumptions, values, concepts, orientations, and priorities" frame research questions, shape analyses, and determine research instruments.[37] Research is driven by desires for Maori autonomy and "cultural well-being," and such research is key to exercising sovereignty and to restoring and building indigenous communities.[38] If research helped subjugate indigenous peoples, then empirical observation and the gathering of data can help liberate them, too.[39] This is precisely the idea that leads some Native American communities to embrace health-related genetic and environmental science and technologies in particular.

Yet Smith also explains that the decolonizing approach of Kaupapa Maori is not synonymous with Maori knowledge and epistemology. Rather, Kaupapa Maori "implies a way of framing and structuring," a set of methods developed by Maori who work in conversation between academic and indigenous communities.[40] Smith concludes her book with a list of "twenty-five indigenous projects" taking place both inside and outside the academy that exhibit the core assumptions she outlines. Such research (re)claims and supports the return of indigenous rights, lands, and histories; builds indigenous archives by documenting oral narratives; and celebrates indigenous survival rather than foretelling the demise of indigenous groups. Decolonizing research—whether or not it also employs indigenous knowledges and methods—reframes views of indigenous peoples, histories, and futures to promote indigenous thriving while also reframing the view of the nonindigenous world.[41]

A similar concept, tribally driven participatory research (TDPR), provides insight into the potential for indigenously governed genetic research. Patricia Mariella and colleagues assert that "indigenous peoples throughout the world have conducted research for millennia; in fact, indigenous knowledge gained by observation and experimentation produced much of the world's foodstuffs as well as many medicines that researchers today seek to assess."[42] CBPR, with its focus on indigenous "participation," is more likely to be initiated and sometimes even driven by those outside the tribe. Mariella et al. describe how tribal government agencies and institutions increasingly perform their own research and maintain the power to invite university or industry collaborators to participate. Puneet Sahota documents one tribe's partnership with a genetics-research institute, both as an investor and as a research partner in studying diabetes and other diseases in tribal populations.[43] Tribal governments also regulate research by approving and denying protocols, publications, and research contracts. TDPR, like Kaupapa Maori, serves indigenous priorities by advocating research as key to the expansion of indigenous governance and sovereignty while not claiming to be an indigenous epistemology or knowledge per se.

At a more fundamental level, we might also ask how "research" conceived as an act in and of itself can disrupt other necessary work[44] and reconstitute what counts as indigenous ways of knowing. For example, tribes in the United States increasingly seek technical assistance from agencies such as the Indian Health Service (IHS) to set up their own research-review boards to govern health-related research, including genetics research. But the process of setting up a research-review board is just the beginning. Staffing the board, hiring outside experts, and training tribal experts in various scientific fields divert community members' energies from other work. Exercising agency in any scientific research process requires training, institution building, and practices that take one off the land and into the university, the conference room, the state agency, and other nonindigenous spaces. Ironically, building bureaucracies and becoming expert in nonindigenous scientific fields is done to protect the very ways of knowing that community members may no longer engage in because their energies are taken up elsewhere. Anthropologist Paul Nadasdy calls attention to this paradox—the development of bureaucratic knowledge in order to save traditional practices that are then not engaged in while the First Nation bureaucracy is being built. In his description of Kluane First Nation ways of knowing, learning is through doing, through the labor that supports Kluane lives. Unlike "research," where knowledge is learned at a distance, "knowing" is not separate from doing and living.[45] Knowledge production takes finite labor power. Research has a cost as well as a benefit.

Feminist and Indigenous Epistemologies at One Table

Like decolonizing methods and indigenous epistemologies that theorize practices and histories of knowledge production about and by indigenes, feminist epistemologies are concerned with knowledge about and by subordinated subjects. Feminists are concerned with the lives of women but do not limit their focus to women. Both feminist and indigenous epistemologists call out the sciences that do not account for their partiality and for representing their views as universal and objective, or value-neutral. Although indigenous and feminist thinkers don't necessarily rely on the same analytical frameworks (for example, indigenous sovereignty infuses indigenous analyses), the two intellectual worlds both push the sciences to be more accountable to the worlds (within which) they study. To this effort, feminist scholars contribute concepts such as situated knowledges, standpoint, and "speaking with, not for" that are helpful for understanding the ethics embodied in this account.

"Objectivity" and Situated Knowledges

Indigenous studies scholar Laurie Anne Whitt explains that those working within a framework of "value neutrality" and "fact-value duality" (often called "positivists") hold knowledge of the natural world to be self-evident. When such knowledge is eventually uncovered or "discovered," it is held to be value-free because nature itself is neutral.[46] Such understandings of knowledge production and notions of truth are critiqued by Donna Haraway, Sandra Harding, and others as conditioned by "the view of everything from nowhere (and everywhere at the same time)." Haraway refers to this as the "God trick," a play on God's omniscience. He (naturally)—the divine Creator—just *is*. God is not situated. Empiricists who claim objectivity and neutrality, in effect, claim a view from nowhere.[47]

Haraway is frequently cited for her conceptualization of "feminist objectivity" in the form of "situated knowledges."[48] What each and every one of us has access to are partial knowledges, because our knowledges are produced within historical, social, value-laden, and technological contexts. The concept of situated knowledges has traveled widely to inform geographical, sociological, and anthropological studies of multiple scientific disciplines.[49] Situated knowledges do not dispose of objectivity. They seek to engage the strengths of both empiricism and constructionism and to avoid their weaknesses.

Haraway is in good company (not always explicitly feminist) in this project. STS scholar Sheila Jasanoff and her students foreground the concept of "co-production of natural and social orders," which I rename "co-constitution" in a small semantic tweak I probably inherited from Haraway. For me, this term avoids the overly constructionist tone of "production." Haraway and her students work consistently in a co-constitutional idiom that views both "natural" and "social" orders as mutually constitutive. Jasanoff adds to the feminist project greater focus on "the authority of the state" in productions of science, technology, and power.[50]

Standpoint

In order to understand the difference that feminist epistemology makes for understanding the emergence of Native American DNA as an object conceived in the work process of particular parties (scientists, genealogists, ancestry-testing companies) and not that of others (indigenous nations), we must pay attention to materiality and discourse while also being attuned to location, multiplicity, and power. We need to be promiscuous in our accounting of standpoints. Haraway calls us to see from multiple standpoints at once, because such a "double vision" is more rigorous. It reveals "both dominations and possibilities unimaginable from a single standpoint."[51] For feminist epistemologists, rigorous, more "strongly objective" inquiry not only does *not* require "point-of-viewlessness,"[52] it actively incorporates knowledges from multiple locations. Rigorous inquiry must also include beginning from the lives of the *marginal*, for example, from the lives of "women and traditional cultures."[53] This is not just a multicultural gesture to pay greater attention from without, it is a call to begin from within, to be *driven* to inquire from within the needs and priorities articulated in marginal spaces. Shifting the gaze sometimes requires new eyes. Sometimes it requires shifting one's feet.

Feminist philosopher of science Sandra Harding explains that feminist standpoint theory is concerned with "the view from women's lives" as a standpoint from which to begin inquiry. She wants women's situations and those of other marginalized peoples in a society stratified by gender, class, race, sexual orientation, and other factors to not be written out of scientific accounts as "bias." Rather, the views from such lives can produce empirically more accurate and theoretically richer explanations than conventional research that treats the views from some lives and not others as bias. Harding explains that the modern/traditional binary that continues to shape both social- and natural-scientific research, as well as philosophy and public policy, "typically treats the needs and desires of women and traditional cultures as irrational, incomprehensible, and irrelevant—or even a powerful obstacle to ideas and strategies for social progress."[54]

The table conversation between feminists and indigenous critics of technoscience should be obvious. Both are "valuable 'strangers' to the social order" who bring a "combination of nearness and remoteness, concern and indifference that are [contrary to positivist thought] central to *maximizing* objectivity." The outsider sees patterns of belief or behavior that are hard for the "natives" (in this case, the scientists), whose ways of living and thinking fit "too closely the dominant institutions and conceptual schemes," to see.[55] In order to precisely represent and effectively confront power, standpoint theorists also pay attention to the fact that subjectivities and lives lived at the intersections of multiple systems of domination become complex. For example, they avoid claiming a single or universal "women's experience"—another reason why their theorizing is beneficial for doing indigenous standpoint theory in the twenty-first century. The feminists recognize that individuals can be oppressed in some situations and in relation to some groups, while being privileged in other instances.

Thus, feminist objectivity helps shape an account of Native American DNA that is critical but that does not simply invert nature/culture, science/culture, and modernity/tradition binaries as we try to see things from an indigenous standpoint. For example, this book does not argue that only indigenous people can speak whereas scientists have no legitimate ground to speak. It does not argue that only "indigenous cultures" or "traditions" matter in circumscribing what it is to be Native American. Nor do federal and Indian policies alone matter. This account is not so naive. Rather, it argues that when we look from feminist and indigenous standpoints, we become more attuned to the particular histories of privilege and denial out of which the concept of Native American DNA has emerged. We might then more rigorously argue against the misrepresentation of this molecular object as apolitical fact that ignores all that indigenous people have suffered and lost in its constitution. The hope is that with greater insight, we might find ways to take more responsibility for the everyday effects, both material and psychic for indigenous peoples and others, of this powerful object and its sister objects—African, Asian, and European DNAs.

Researching, Consuming, and Capitalizing on Native American DNA

Three chapters of this book explore the use of Native American DNA by genetics researchers, consumers, and for-profit companies. Although no chapter focuses on Native American tribal uses of DNA testing, tribal engagement with DNA is implicated especially in chapter 2. In the forward-looking conclusion, I also emphasize indigenous governance, broadly speaking, of genetic science as a necessary intervention, both for building indigenous capacity to govern and for democratizing scientific practice.

In chapter 1, "Racial Science, Blood, and DNA," I explore the concept of blood by drawing on key texts from history and anthropology that treat the subject in both global and U.S. terms. In the latter case, I compare dominant U.S. concepts of blood with Native American or "tribal" blood concepts by combining insights from the aforementioned literatures with those from legal scholarship on the relation of blood to the conceptualization of the Native American "tribe" in nineteenth- and twentieth-century U.S. law. One cannot talk about blood in a Native American context without exploring its co-constitution with the concepts of tribe and race in the colonial practices of the United States. In addition, I look at how "race" and "tribe" organize Native American identity in different, sometimes overlapping ways and the looping implications for U.S. concepts of race broadly. I build on the valuable work of established race scholars and, in particular, younger race scholars who study how Native American race has been conceived to reinforce the division between white and black. Such conceptions facilitate ownership claims to Native American history and cultural patrimony by the white nation. I add an analysis of how genetic concepts further support the ownership of Native American history, bodies, molecules, and identities by whites.

In chapter 2, "The DNA Dot-com: Selling Ancestry," I analyze the technical and cultural production of DNA-testing companies that target Native American ancestry and identity, especially their overly simplistic claims about the correlation between racial and ethnic identity and genetics. The chapter is primarily textual analysis of company Web pages, print advertisements, trade-show advertising materials, company representative statements and interviews in the popular press, and correspondence with company representatives. I also draw on participant observation, for example, attendance at national tribal-enrollment conferences for tribal staff where DNA-testing companies both give technical talks and advertise. In the chapter, I focus on a half dozen companies and their practices that together represent the array of technologies offered for ascertaining Native American ancestry. Some of these companies, because of marketplace dynamics, are no longer in business, but the products and texts generated by the companies remain technically and culturally current. Companies that bring different techniques of analysis to bear on Native American identity treat differently the concepts of "tribe" and "race." Some exhibit more understanding than others of the overlap and contradictions between those categories, especially as they relate to tribal sovereignty or indigenous governance. This chapter explains why a turn to genetic tests will not solve the intractable problems associated with blood-quantum and other nonscientific tribal-enrollment policies despite company promises of scientific precision.

In chapter 3, "Genetic Genealogy Online," I explore the practice of "genetic genealogy," or using ancestry-DNA tests to fill in documentary gaps that arise in "family tree" research. Doing family genealogies is a top American pastime, and genetic testing is a quickly growing practice among genealogists. Whereas chapter 2 highlights the activities, texts, and claims of DNA-testing companies, chapter 3 is based on participant observation and focuses on the activities, texts, and claims of a group of DNA-test consumers who were active in one online community in the mid-2000s, a Listserv dedicated to sharing genetic-testing information and to mentoring among genealogy researchers. The chapter also draws an analysis of tens of thousands of archived posts from 2000 (when the list was founded) to 2005. In this chapter, I am especially attuned to politics of race and property as they play out between the categories of "Native American" and "white" ("Anglo-Saxon," "WASP," "European American," "Caucasian," and so on) on this Listserv, which is dominated by largely self-identified whites.

Chapter 4, "The Genographic Project: The Business of Research and Representation," treats the politics of race and indigeneity as expressed in the Genographic Project and by Spencer Wells, National Geographic explorer-in-residence and Genographic's principal investigator. Launched in 2005, Genographic aims to trace the global migratory history of humans (pre-1492, it is implied) by sampling one hundred thousand "indigenous and traditional peoples" around the globe. This chapter performs a textual analysis of Genographic public relations and research output—Web sites, videos, news articles, press interviews with project organizers, and research papers—that keep alive five problematic narratives at the intersections of race and science: that "we are all African," that "genetic science can end racism," that "indigenous peoples are vanishing," that "we are all related," and that Genographic "collaborates" with indigenous peoples. Although Genographic might seem to liberate "genetics" and the "population" from their older counterparts "blood" and "race," the project actually conjoins older racial ideas and even racist practices with newer discourses of multiculturalism and the idea that "we are all related." This chapter pays special attention to the implications for indigenous identities, indigenous knowledges, and indigenous governance of the Genographic Project specifically, but it also highlights the broader colonial context—how Genographic relies on older notions of race while constructing identities and histories as disproportionately genetic.

In the Conclusion, "Indigenous and Genetic Governance and Knowledge," I look toward a more hopeful future for the interactions between genome science and indigenous peoples. My central role in this volume is as cultural analyst and critic. But I was also trained as a community and environmental planner, and I grew up under the moral and political tutelage of tribal and urban Native American community planners, institution builders, and activists. I call attention to the critical and ethical lapses of genome science throughout the following chapters, but I am also moved to foreground promising developments that might take us beyond the present status quo. I begin the Conclusion with an overview of key points from Canadian genetic scientist Roderick McInnes's 2010 American Society of Human Genetics (ASHG) presidential address in Washington, D.C.,[56] in which he encouraged geneticists to acknowledge indigenous "intercultural frameworks"[57] that can help genetic researchers better respect indigenous claims related to property and sacredness of biological materials. McInnes highlights the efforts of the Canadian Institutes of Health Research (CIHR)—the equivalent of the U.S. National Institutes of Health (NIH)—to use an intercultural framework to guide genome research on Aboriginal peoples in Canada. The respect for Aboriginal sovereignty and the collaborative spirit that McInnes advocates are in sync with changes to standard research and institutional practices suggested by indigenous critics, bioethicists, and social-science scholars in the United States—suggestions that have met skepticism from geneticists. Issued instead from the ASHG presidential pulpit, perhaps such ideas will be more seriously considered by the mainstream of U.S. genome science.

The power inequities are real in this world in which DNA narratives increasingly grab center stage in the telling of human history and the construction of human identities. But those relations of power are not as linear as they used to be. As indigenous peoples push back on those who gaze on them and would extract their biological and cultural resources, and as those who do science become more diverse, the sciences are not only a culprit, they are a site for change.

1

RACIAL SCIENCE, BLOOD, AND DNA

Blood metaphors have a primal quality; those who use them, even today, assume their meaning is understood. Now they imply a genetic element, but this was not always so.

—Melissa Meyer, *Thicker Than Water: The Origins of Blood as Symbol and Ritual*

"We used to think our fate was in our stars. Now we know, in large measure, our fate is in our genes."

—James Watson, quoted by Leon Jaroff in "The Gene Hunt," *Time Magazine*

The phenomenon of Native American DNA can be understood in all of its richness only if it is understood as co-constituted with U.S. race categories, which themselves are coproduced with Euro-American colonial practices, including eighteenth- through twentieth-century U.S. race laws, policy, and programs. The meanings of Native American DNA and the practices that in part produce it must also be articulated with older meanings of "blood" as the substance of inheritance in pre-genomic eras. There are very particular articulations of blood and gene metaphors, politics, and materialities with respect to Native American race identity and tribal citizenship. This chapter presents "genealogies" or "redescriptions" of the often unintended relations and conjunctures of the scientific and other social practices that have produced this technoscientific object, Native American DNA, and that continue to give and challenge its meaning.[1]

Following Foucault, I do not treat Native American DNA as the latest moment in a linear, progressive, or purposeful history of Native

American identity. Instead, I pursue a genealogy of the concept, focusing on the accidents and deviations in the history, the way that "truths" are produced out of disunity, and the way DNA reflects power as an object identity and knowledge.[2] In other words, particular strings of molecules in particular bodies are not in any way simply a transparent reflection or indicator of a particular genetic population. The concept of Native American DNA is instead constituted of relations between molecules, happenings, instruments, and minds. It is enabled by preexisting and overly discrete notions of genetic difference between groups. It loops back then to shore up genetic understandings of difference, giving credibility and authority to a particular way of knowing and seeing.

In the next section, I briefly discuss how the human sciences, mostly in Great Britain and the United States, conceived of "race" as an object of inquiry during the nineteenth and early twentieth centuries. This subject is extensively treated elsewhere,[3] but revisiting some of the highlights of that era of racial science will help us to see more clearly the complex social-material nature of Native American DNA.

A Note on Native American "Race"

In addition to molecular concepts, this chapter treats the role of "blood politics" in both the human biological sciences and U.S. colonial policy in the nineteenth and twentieth centuries.[4] Blood politics are critical to understanding the constitution of the U.S. "tribe" as a particular social-political entity. Blood, both alone and, more recently, as it has been entangled with molecular concepts, also conditions this notion of Native American race. Tribal folk often quibble with those who would refer to them/us as a racial group: "We are not a race, we are tribal citizens. Citizenship is different from race." Yes and no. We Native Americans have been racialized as such within the broader American cultural milieu. We privilege our rights and identities as citizens of tribal nations for good reason: citizenship is key to sovereignty, which is key to maintaining our land bases. But race has also been imposed upon us. Race politics over the centuries in both Europe and the United States have conditioned our experiences and opportunities, including the federal-tribal relationship. They have impinged upon our ability as indigenous peoples to exercise self-governance.

Race Science in the Nineteenth and Early Twentieth Centuries

The nineteenth century has been characterized as an era in which earlier ideas of different races solidified into the belief in "racial type."[5] Arguments about race articulated in the sciences spawned associated racist discourses that have been dubbed "scientific racism." "Racial science," by contrast, can be used to refer to the scientific investigation of "racial differentiation, the mere discrimination *between* races and their purported members," which, as race theorist David Theo Goldberg points out, is not necessarily a racist act. However, classification of races or "human groupings on the basis of natural characteristics" occurred within nineteenth-century anthropology that assumed that racial "differentia [are] adequate to establish the laws of behavior for [racial] members." Although racial differentiation is not inherently racist, it "begins to define otherness, and discrimination *against* the racially defined other becomes at once exclusion of the different."[6] Thus, race science and scientific racism, though different in important ways, are linked.

Our contemporary understandings of race in the United States, including within the field of human genome diversity research, emerged in important ways out of racial science and related racist discourses in the nineteenth century. Anthropologist and historian of science George Stocking, in his classic *Race, Culture, and Evolution* (1968), examines what he calls "survivals" within changing patterns of racial thought. Akin to James Clifford's notion of "articulations," survivals bear "clear logical if not always easily establishable lineal relations" to earlier positions on race.[7] Stocking describes race as a scientific object of inquiry and traces its changes over time, with new and old elements, bodies of knowledge, and techniques or technologies of inquiry being conjoined and disarticulated. His archival work shows how scientists throughout the nineteenth century foregrounded ideas of biology as identity and scrutinized physical differences between human beings. Races became "race," and "race" came to be understood as constituted by biology. It was then that racial science spurred scientific racism. Rather than individuals simply being classified into different races, race itself became the chief explanation for observed human differences, "the permanent inherited physical differences which distinguish human groups."[8]

It has been argued that science led popular opinion and, alternatively, that popular ideas led science to formulate ideas about race.[9] A coproduction framework allows for both assessments to be correct. Science and broader societal beliefs embed one another and come into being unevenly together. We need to keep this coproduction process in mind as we look at the emergence of Native American DNA later in the chapter.

One Human Race or Multiple Human Species?

In an essay analyzing nineteenth-century polygenism (the notion that the different races had separate origins, namely, separate creation events) as the first biological theory to support the idea of the "intrinsic inferiority [of] despised groups," Stephen Jay Gould describes what I would call a coproduction process at work. Not only did nineteenth-century racial science inform societal thinking about race, but also popular ideas about heredity and hierarchy—the pervasive racial attitudes of Americans in the nineteenth century—informed the pursuit of scientific knowledge. Such attitudes, whether held by "hard-liners" (polygenists) or "soft-liners" (monogenists, the liberals of the day), look today like embarrassing racist ideas. Liberal abolitionist Charles Darwin held that "negroes" and "Australians" (seen in the day as members of a broader "Negro Group" classification) were intermediates between Caucasians and apes.[10] But he was liberal in that he saw all human races as belonging at least to the same species. The argument about whether different races constituted different species was perhaps the most important racial science debate of the nineteenth century.

The work of two prominent polygenists, Louis Agassiz and Samuel Morton, illustrates how popular stereotypes were entangled with the science of the day. Agassiz's taxonomic work focused on race ranking and on judging the characters of races. In line with accepted stereotypes that informed U.S. policy, Agassiz saw Indians as courageous and proud, more likely to go to their deaths in battle than to submit to U.S. dominance. Therefore, the Indian problem would be solved as the Indian vanished. By contrast, Agassiz saw blacks as submissive and obsequious, reflecting the dominant view of the "pliable Negro" that might be a permanent presence and problem within the nation.

Scientists today study and measure molecular sequences and frequencies in people from different parts of the world in order to classify both markers and individuals who possess such markers as belonging to particular "populations" or "(biogeographical) ancestry" categories (these in lieu of the older "race" categories).[11] But in the nineteenth century, studying and measuring skull and bone sizes and shapes were state of the art in the study of race, what biological anthropologist–cum–historian of science Jonathan Marks calls a "radical materialism." Like genetic markers in the twenty-first century (although we speak more about "probability" today), morphological data was used both to delineate the biophysical markers of races and to categorize individuals and their body parts within those categories. In the nineteenth century, it was a widespread belief that the brains of different races were inherently different, contributing to the production of different civilizations, indeed different levels of civilization, and thus guaranteed different potentials in the different races. Both Morton and another polygenist studier of skulls, Josiah Nott, "without adjusting for age, sex, body size, or nutritional status of their specimens, invariably found the brains of Europeans to be larger than those of other peoples, thus explaining the widespread subjugation of the latter by the former."[12]

Published in 1859, Darwin's *On the Origin of Species* asserted that all humans had a common ancestor. However, unlike other monogenist thought, Darwinian evolution through natural selection placed that ancestor in the deep past, far beyond the temporal range of the garden of Eden. Shockingly, Darwin also suggested that humankind's ancestor was nonhuman. Although Darwin technically ended the debate about there being multiple human origins, he and others remained influenced by polygenist thought. Limited to the fossil record for evidence (and, let us not forget, with prefigured notions of discrete biological boundaries and hierarchies), Darwin and others continued to view races as fixed and discrete. Darwin's new evolutionary narrative proposed that shared ancestors lived so long ago that humans were no longer subject to environmental influence—that natural selection had in antiquity fixed the physical and dependent cultural capacities of the different races.[13]

Race Hierarchy or Cultural Relativism?

After the end of the monogenist-polygenist debate, a second related debate emerged in the human sciences: that between race hierarchy and the idea of racial or cultural difference. After Darwin, the business of ranking races continued but was taken up through an evolutionary framework. Darwin emphasized evolutionary continuity and a slow human evolution from lower animals. Social Darwinists subsequently theorized that racial disparities, their causes, and their solutions could be explained by evolution through natural selection. In the struggle between races, the fittest had triumphed over barbarism to produce European civilization.

At the turn of the twentieth century, the landscape of racial thought shifted profoundly as Franz Boas (and later his students) advocated what is now a mainstream idea—that "cultures" are plural and relativistic, and that they belong to all societies.[14] This as opposed to the older humanist concept of culture that social Darwinists adhered to, in which culture is seen to consist of art, science, and knowledge in absolute and singular terms possessed in varying degrees by more- and less-civilized peoples. Social Darwinists held that "culture" is acquired progressively through evolution, freeing those who have it from the controlling forces of nature, reflex, and "traditions" associated with lower evolutionary status.[15] Boas developed his oppositional notion of cultures (plural) through field study of the practices of North American Natives. Group cultural phenomena, he proposed, derived from "specific and complex historical processes" rather than signifying earlier stages in the racial hierarchy of humanity. The evolutionary superiority attributed to Western civilization was not provable. Such views came from proponents whose actions were conditioned since birth by Western civilizational values.[16] Not unlike Charles Darwin, Franz Boas was a progressive thinker for his era. And not unlike Darwin, he conjoined newer, more liberatory ideas with older race thinking. Although Boas generally opposed the idea of simple biological determinism, he substituted "culture" for biological heredity as a determinant of mental differences between races and studied the influence of culture on racial formation. Because of the influence of Boas and his students, the social sciences moved away from biological explanations of race.

In the biophysical sciences, Boas's culturalist views were not so influential. Other factors explain the downplaying of race as a biologically valid category in those fields. Although the nature-versus-nurture debate helped dampen the enthusiasm for race as an object of biological inquiry, perhaps more important is that race proved technically elusive to materialist scientists. Despite ever more precise racial classification methods and tools (such as those related to skulls and eventually to the classification of blood types), pegging discrete racial categories proved impossible. Categories such as ethnicity and nation also muddied scientists' clean delineations of race.[17] As a result, many in the biophysical sciences backed away from race as a primary object of inquiry. An opening panel of the American Anthropological Association museum exhibit *RACE: Are We So Different?* which is touring science museums nationally at this writing and which includes human genome diversity content, makes the oft-repeated claim that scientists agree that race is socially constructed.[18] Others argue that the science of human genome diversity research has disproven the existence of race and that therefore genomics should end racism.

Yet some scientists continue to find biological meaning in race into the twenty-first century. They continue to look for and assert biological markers of race.[19] The next chapter, on DNA-testing companies, brings this clearly to light. Jenny Reardon contends that the life sciences continued to be "a major site for the construction and reconstruction of race following World War II."[20]

Eugenics

A genealogy of eugenics reveals a pattern of advance, retreat, and reconfiguration of biologically based ideas about race and their ugly associations with power. The concept of eugenics was introduced in 1883 by Sir Francis Galton, Charles Darwin's cousin. Galton promoted scientific research and development of social policy that aimed to improve the racial pool of humans through selective breeding.[21] The eugenics movement survived into the twentieth century and was particularly well organized in Great Britain, the United States, and Germany. Scientists and other eugenicists from all three countries cooperated through mutually supportive research networks. Germany, in legislating the behavior and right to life of Jews, homosexuals, the disabled, those convicted of crimes, and others, looked to the United States for guidance on laws related to sterilization, immigration, and miscegenation. Recent immigrants (with lower rates of literacy) were targeted because of lower IQ scores. In the United States, immigration restrictions were targeted at "Jews, Poles, and Southern Europeans while enabling immigration of people of British and Northern European descent."[22] Laws forbidding sexual relations and marriage between races were also enacted to maintain the purity of the American (white) population. As late as 1967, sixteen states still had laws forbidding interracial marriage.

After World War II and Nazi Germany's genocidal eugenics, eugenicists in other countries worked to separate themselves from the policies and tactics of Germany. Simultaneously, the critiques of Boas and late nineteenth-century sociological work that emphasized the role of environment in shaping human behavior[23] came more into the mainstream, aiding a transition away from older incarnations of biological studies of race. Newer incarnations, however, would rise in their place.

"Population" and Race

In 1900, the science of genetics was rediscovered. Gregor Mendel's obscure mid-nineteenth-century studies on pea plants and the heritability of physical characteristics were brought to light at a meeting of the German Botanical Society. Although the term *gene* was coined several years after *genetics* (in 1906), there was no consensus on a gene's actual material reality. There were vague references in the literature to such things as "elements" within gametes that had something to do with the characteristics of organisms. Some scientists talked of a gene being "the fundamental unit of heredity," a vague definition that persists to this day in textbooks.[24] But beyond that, little was known.

In the 1930s, a new science, a "populational, genetical science of human diversity," emerged.[25] It was nourished by the decreasing viability of racial science's theories, techniques, and propositions, by the renunciation of the old science by younger scientists, and by the cultural work of social scientists. The racial horrors of WWII dealt the old race science a hefty blow. The new science found attempts to classify humans in a zoological manner irrelevant. Genetics at this stage focused on studying how human groups "adapt, how they vary, and what the impact of their histories has been upon their biology."[26] Rather than race being shunned, both physical anthropologists and geneticists regarded race as an important factor in the study of human variation and evolution, but they redefined it, "in the wake of scientific and political developments," as "population."[27]

Population genetics has infused anthropological practices, resulting in the subfield of molecular anthropology, which studies human beings, their movements, their relatedness, and their so-called origins. Jonathan Marks describes molecular anthropology (a name coined in 1962) as the application of biochemistry to "classically anthropological questions." He argues that the field weds up-to-the-minute genetic technologies with the "folk knowledge" of anthropology. Molecular anthropology is technology-driven. Thus, he laments, anyone—regardless of how much anthropology they know—can do anthropology. This has produced research interpretations that are what Marks provocatively calls strikingly culturally naive.[28]

For example, the central dogma of molecular biology introduced by Francis Crick in 1958 holds that DNA contains the complete genetic information that defines the structure and function of an organism, and that the flow of genetic information is DNA to RNA to protein. Punning the central dogma, Marks describes a "reduction of the important things in life to genetics as the Central Fallacy of Molecular Anthropology."[29] Evelyn Fox Keller has also called attention to the central dogma's embedded cultural expectations. She elaborates that the "crucial point of the central dogma is its insistence on unidirectional causality, its repudiation of the possibility of a substantive influence on genes, either from their external or from their intra- or intercellular environment. Instead of circular feedback, it promised a linear structure of causal influence, from the central office of DNA to the outlying subsidiaries of the protein factory."[30] In essence, "DNA makes RNA, RNA makes proteins, and proteins make us."[31] Since the 1950s, the effects of this conception of the linear influence of genes on our concepts of identity have been profound for scientists, and for all of us.

"Native American DNA": Basic Science and the Tests That Define It

In Indian Country, we are not as steeped in gene talk when we speak of kinship and identity as is the broader American public. We remain profoundly influenced by the language of blood, which I address in the final section of this chapter. But things are changing rapidly, especially now that some gaming tribes make extensive use of DNA testing. For my tribal audience in particular, a review of the basic science of DNA testing is in order. Without that, we in Indian Country will find it difficult to accurately assess the risks and benefits to tribal sovereignty of our increased exposure to genetic tests and to genetic research. Indian Country will remain largely at the mercy of non-Native technical advisers who tend to work for the very DNA-testing companies or research institutions that stand to benefit from the geneticization of Native American identity.

What Is DNA?

Four nucleotide bases, or chemicals, make up our deoxyribonucleic acid, or DNA. They are adenine (A), cytosine (C), guanine (G), and thymine (T). An organism's total DNA makes up its "genome." Within a genome, the bases are ordered in many different combinations. All humans, of course, share the same DNA structure and thus basically the same genome, but we differ individually at certain places. For example, the order of bases—more commonly called the "sequence"—can differ at particular locations from person to person. Such differences are most often inherited, although sometimes they arise through spontaneous mutation.

The majority of our DNA is within our chromosomes. Humans normally possess forty-six chromosomes, very long strings of DNA packaged with proteins, in the nuclei of our cells. Our chromosomes come in pairs, one from each parent, numbered 1–22. The "sex chromosomes" constitute the twenty-third pair. Genetic males possess one Y chromosome, inherited from the father, and one X chromosome, inherited from the mother, whereas genetic females possess two X chromosomes, one from the mother and one from the father.

We also have DNA within our mitochondria (mtDNA). The mitochondrion is a structure, or "organelle," found within the cell's cytoplasm that generates power for cellular functions by converting energy from food molecules. Mitochondria are often described as the "powerhouses" of cells. Each cell has from a few hundred to more than one thousand mitochondria (except for red blood cells, which have none). Mitochondria carry their own DNA, thus giving us evidence that they were once independent organisms, probably the descendants of bacteria that lived billions of years ago outside of cells and were later incorporated into cells. MtDNA is especially useful for gauging biological relatedness on the maternal line. Unlike nuclear DNA, the sperm does not contribute mtDNA to the zygote during fertilization, because it contributes little cytoplasm, where mitochondria are found, to the zygote. Because there is no recombination of maternal and paternal DNA in this case, we all inherit our mtDNA solely from our mothers, and they from their mothers, and so on.[32]

DNA tests reveal particular orderings of nucleotide bases in our genomes. In relation to the known patterns of other individuals and populations, such orderings, or sequences, can tell us to whom we are most probably genetically related. The DNA profile, or fingerprint, can tell us to whom we are closely related, including mother, father, sister, brother, aunt, uncle, niece, nephew, and grandparents. Or the genetic-ancestry test can tell us from which male and female "founders" we are descended and on which continents those ancestors probably resided.

Given the prevalence of genetic determinism in our popular culture, it seems important to clarify that most markers used to show ancestry and close biological relationships are not known to code for anything. Previously designated "junk DNA," noncoding DNA does not comprise genes. Genes are, by definition, nucleotide sequences that, via RNA, encode proteins that in turn produce or contribute to certain traits. It is estimated that more than 98 percent of our DNA is noncoding.[33]

Mitochondrial DNA and Y-Chromosome Lineage Tests

Direct-to-consumer (DTC) DNA-testing companies offer three types of genetic-ancestry tests. MtDNA and Y-chromosome tests look for DNA markers that signal a person's descent from "founding populations" that inhabited regions and continents of the earth thousands of years ago. Scientists call these markers, for example, "Native American," "European," "Asian," and "African" "haplogroups," lines of descent dubbed using capital letters: A, B, C, D, and so on. Within haplogroups, there are further subtypes called "haplotypes." The first type of test examines DNAs in the mitochondria, which are inherited solely through the maternal line. The second type of test examines DNAs in the male-specific region of the Y chromosome, which does not recombine. Because they do not recombine, mtDNA and Y markers indicate clearly maternal and paternal ancestral lineages.[34] Most ancestry DNA tests for sale in the online marketplace are of this variety.

Commercial DNA tests generally involve a physically noninvasive procedure. After rubbing a sterile cotton swab on the inside of the cheek and sending it in a vial to a testing company, an individual will be told usually within weeks what the particular markers examined reveal about their biological ancestry. Ancestry DNA tracing that takes place as part of academic research often involves a researcher drawing blood from a subject, which provides more DNA for use in research.

Native American Maternal Lineages

Within the mtDNA, scientists have found certain nucleotide sequences that they call "Native American markers" because they are understood to have been inherited through the generations from genetic females within populations that first settled the "New World." On the mtDNA, five Native American haplogroups are commonly referred to. These are clusters of closely linked markers inherited together. All five haplogroups have been identified by research scientists in "prehistoric Native North American samples," and it is commonly asserted that most living Native Americans possess one of the mtDNA haplogroup markers.[35] Of course, most living Native Americans have not had their DNA studied, and the definition of who is Native American for the purpose of sampling is not always in line with broader social and political definitions of who is Native American. Thus, mtDNA testing can be misleading, and it is premature to state that "most" native peoples possess such DNA. Subsequent chapters address these problems.

The so-called Native American haplogroups are labeled A, B, C, D, and X. A through D are traced to Asia, but X is found in Europe and hardly ever today in Asia except in the Lake Baikal region of southeast Siberia. It is believed that those with X markers in Europe and the Americas have common ancient ancestry in the Lake Baikal region. Some speculate that X in the Americas may trace to pre-Columbian migrants from Europe (rather than to common ancestors), but that assertion is controversial.

Native American Paternal Lineages

Genetic males can also have their Y chromosome examined for Native American markers. Genetic females do not possess a Y chromosome. They often find male relatives in their genetic father's line to have tested if they wish to know about these lineages. The major Native American haplogroups found on the Y chromosome are labeled C and Q and are thought to trace probably to southwest Siberia.[36] Within those haplogroups, the lineages C-3b and Q-M3 (also referred to as Q1a3a) are the most commonly found Y lineages among Native American populations sampled.[37]

Interestingly, an important 2004 paper by Stephen Zegura and colleagues explains that three major haplogroups, C, Q, and R, account for "nearly 96% of Native American Y chromosomes" (based on a sample of 588 Native Americans). But R is not considered a "Native American haplogroup," because it probably reflects "admixture" with Europeans. A good proportion of the Native American males sampled had Y haplogroup markers that derived from other, more recent ancestry.[38]

The shortcomings of these tests are that they examine a very few lineages that comprise a very small percentage of one's total ancestry, less than 1 percent of total DNA. Going back only ten generations, each of us has about one thousand ancestors. One could have up to two Native American grandparents and show no sign of Native American ancestry. For example, a genetic male could have a maternal grandfather (from whom he did not inherit his Y chromosome) and a paternal grandmother (from whom he did not inherit his mtDNA) who were descended from Native American founders, but mtDNA and Y-chromosome analyses would not detect them.

The Autosomal Marker Test

There are two additional genetic tests that circulate with discourses of Native American identity. First is a third type of genetic-ancestry test, an "autosomal" test, which looks for markers across the part of the Y that recombines and all of the other chromosomes. Autosomal markers are inherited from both parents, and the vast majority of our DNA is autosomal. Given this, autosomal tests fill in some of the gaps left by mtDNA and Y-chromosome analyses. They trace the DNA contributions of many more ancestors, albeit with much less clarity as to where in one's lineage those ancestors occurred. In the next chapter, I discuss a prominent test for these markers, the test formerly known as AncestrybyDNA™, patented by the former DNAPrint and now sold by DNA Diagnostics Center.

In basic terms, autosomal markers are the result of mutations in the DNA of individuals. Those mutations result in single nucleotide changes from one individual to another, called "single nucleotide polymorphisms," or SNPs. Because a child inherits only one member of a chromosome pair from each parent, there is only a 50 percent chance that the SNP appearing on a parent's chromosome is passed on to the offspring, and so on. Thus, many SNPs disappear over the generations, but some do not. In fact, in some populations some SNPs became quite common. Groups of such SNPs are found in high proportions in certain parts of the world and in much lower proportions in other parts of the world. Some companies sell DNA tests that look for combinations of SNPs, because in combination, as opposed to singly, they have a much higher probability of strongly indicating one's ancestry in populations from a certain part of the world. For example, if a particular group of markers has been found to occur in 75 percent of living people sampled in Africa but in only 6 percent of living Europeans sampled and in 2 percent of living Native Americans sampled, such markers are seen as indicating probable ancestry in Africa.[39]

The DNA Profile

A fourth type of DNA test, the "DNA profile," is marketed specifically to federally recognized tribes to aid enrollment decisions. Sometimes also referred to as a "DNA fingerprint," this is the same form of genetic analysis used in criminal cases, for example, to identify an individual associated with a biological material (such as blood, hair, or semen) left at a crime scene. Another common use for this form of analysis is as a paternity test, but it is also of use in proving genetic maternity and other close biological relationships (siblings, cousins, aunts, uncles, and grandparents) with lower but still very high degrees of probability.

The DNA profile is increasingly used by tribes in the United States and by First Nations in Canada. With proof of genetic parentage, some enrollment offices allow a parent's blood-quantum documentation (explained shortly) to be invoked in order to determine an applicant's blood quantum and in turn to process that applicant's enrollment. I further address the parentage test and its implications for Native American governance and identity in the next chapter. I analyze the DNA profile as part of the phenomenon of "Native American DNA." A genetic scientist would not describe parentage analyses as such. But I do it to make a broader point: that the category of Native American DNA ties Native American history, tribal identity, and ideas of race to molecular concepts. DNA profiling does this just as effectively—if not more so—than do genetic-ancestry tests at this historical moment. A crucial question is, How are DNA concepts being articulated with blood concepts that have for much longer been central to configuring Native American identity and citizenship? Familiarity with the basics of DNA is the first step to understanding. Now we need to dive into the complex politics of blood.

Material-Semiotic Blood and the American Racial Imagination

Until just after the inception of the field of human-population genetics and the study of human genome diversity in the 1950s and 1960s,[40] blood in its semiotic richness still structured categories and laws used to manage individual bodies and races. Antimiscegenation laws permeated the American South from the colonial era until 1968, when the race codes of southern states were ruled unconstitutional.[41] It was not genetic science but a storm of mass movements against race oppression in the 1960s that brought an end to race regulations and their language of blood rules.

Blood has become less fashionable and is outside of law as a mechanism for the management of racialized bodies, and genetic metaphors increasingly colonize our vocabularies of self, inheritance, and destiny. Yet blood concepts still inhabit discourses of race and identity in the United States, especially in Native American tribal communities. Since its conception, "Indian blood" has enjoyed a unique place in the American racial imagination, and tribal communities are managed (by others or by us) according to the precise and elaborate symbolics of blood. Considered a property that would hold Indians back on the road to civilization, Indian blood could be diluted over generations through interbreeding with Euro-American populations. Indians were seen as capable of cultural evolution (unlike Africans) and therefore of cultural absorption into the white populace.[42] "Kill the Indian, save the man" was a mantra of nineteenth-century U.S. assimilation policies.[43] Indian blood could also be overcome via mandatory boarding-school education, bans on religious practices, and the destruction of communal living and property arrangements. Through both physical and cultural dilution, the Indian was thought capable of being reconstituted, reeducated, and made into a more fully advanced human.

Paradoxically, Indian blood also became a desired object. Its complete loss would be lamented, as the "First American" is central to the country's nation-building project—to constructing moral legitimacy and a uniquely American identity.[44] The Indians, at the moment of their subdual, become ancestors. Yael Ben-zvi, in one of the best recent papers on racial formation in the United States, writes about the transition after the nineteenth century in "the U.S. racial imagination from a tripartite to a binary model."[45] Red disappears, leaving only black versus white—this despite the demographic explosion among Native Americans in the twentieth century.[46] Ben-zvi links the discourse of presumed Native American disappearance through death or racial absorption to the absorption of Indian property by a white nation and the need to legitimate that transfer of land, cultural patrimony, and even biology.[47] That red became no longer perceived as a racial category independent and distinguishable from white helped consolidate the black–white binary.[48] Indians were seen by scientists and liberal policy makers as capable of moving along the evolutionary continuum of racial hierarchy, closer to the Europeans and, later, white Americans at its pinnacle.[49] Africans were not considered capable of this movement, and African (American) bodies and culture were excluded from the nation.[50]

The disappearance of red; the consolidated black–white binary; and claims to Native American biological, historical, and cultural patrimony permeate the science of human genome diversity, the practices of genetic genealogy, and nineteenth- and twentieth-century rearticulations of the tribe. Genomics has come to the fore of U.S. development and governance strategies. And Native American DNA emerged in the late twentieth century as an observable object for the study of race (reconfigured as population) in concert with studies of human migrations. Throughout this chapter, I weave back and forth between blood and genetic concepts to explain the shifting roles they play in how we perceive the inheritance of Native American biological and cultural patrimony. Blood is no longer a robust scientific object for the rigorous study of race or culture, but knowledge of the intricate arrangements of adenine, guanine, thymine, and cytosine cannot reconfigure the Indian without coming up against the apparatus of Indian blood that is still influential in other quarters.

Federal and tribal government blood rules such as "blood quantum" have helped to constitute unique formations of the Indian and the tribe for more than a century. Where the federal policy project of the nineteenth century was to *de*tribalize,[51] what has happened in effect is a *re*articulated tribalization of Native Americans in blood fractions and through bloodlines. A key mechanism meant to dilute Indians out of existence has reconfigured them and, in the particular form of lineal descent, propagates them. If the material properties of blood—the red fluid itself—are no longer legitimate for the study of race, symbolic blood remains very much at play in twenty-first-century sociopolitical formations of the Indian. The complex semiotics of blood must be understood if we are to gauge the opportunities and barriers for Native American DNA to shift the boundaries of race, tribe, and indigeneity.

For much of human history, actual and symbolic blood permeated discussions about and experiences in the social and political realms. Piero Camporesi writes that in earlier centuries, "barbers, phlebotomists, pork butchers, midwives, brothers hospitalers, opened, closed, cauterized veins with appalling indifference."[52] Blood was let flow for medical and religious reasons, in torturous executions in some lands and eras, in sacrifices in others. Nonhuman animals were butchered and hung in the open, and their blood was drunk raw or cooked. In what is today western Europe, during the rise of the medical sciences, there were live human dissections of criminals sentenced to death. Menstruating women, considered both taboo and powerful, preoccupied peoples around the world. In many cultures, menstruants were secluded, and blood signified contamination of men by women. But menstrual blood also nourished. It represented fecundity and has been used as fertilizer. Blood signified moral and physical purity and social solidarity. The color and condition of blood preoccupied the minds of thinkers and represented the essence of life and personhood. Some believed it to be "the seat of the soul."[53] Jean Dennison, in her ethnography of the 2004–6 Osage Nation citizenship and government-reform process, notes the "real power" of blood for Osage.[54] Melissa Meyer concurs that "many, many peoples . . . have conflated literal red liquidity with the special characteristics of a culture or people. Instead of serving merely as a potent metaphor, the viscous blood flowing through human bodies was believed to convey essential attributes." Symbolized and ritualized across human cultures, blood today "most often connotes lineage, descent, heredity, and race . . . making it easy to forget [without the same bloodletting that ancestors engaged in, Camporesi would add] that real blood is red and fluid."[55]

Against a backdrop in which blood signifies contamination and purity, in which blood can link individuals to or distinguish them from a society or group,[56] race emerged as a tool for the state and for science to categorize and trade in human bodies and body parts.[57] Blood types illustrate the point. When ABO blood types were uncovered, in 1900, they were investigated by scientists as indicative of race. Blood types exhibit the proportional patterns of inheritance demonstrated by Mendel, being caused by a single gene. ABO blood types and the alleles for them occur in higher and lower frequencies in different populations.[58] Nearly all populations exhibit all three blood types, A, B, and O, but proportions of each blood type within different populations differ, sometimes dramatically. Sometimes, populations we consider racially disparate exhibit similar frequencies, and blood groups do not align with morphological characteristics. A 1919 research paper explained that "the Indians [i.e., south Asians], who are looked upon as anthropologically nearest to the Europeans, show the greatest difference from them in the blood properties."[59]

Jonathan Marks explains that "the European cultural mystique about blood and heredity probably aided in the credibility of the serological work—and somewhat uncritically, for it quickly became clear that the claims about the racial study from blood were just insupportable." Marks writes that research that attempted to track race to blood group neglected "the fact that 'blood' is a *metaphor* for heredity, not heredity itself."[60]

Our Ancestors Didn't Know about Genes, and Why That Matters

Anthropologist David Schneider's seminal *American Kinship* (1968) casts light on blood both as a biogenetic substance and as a metaphor for relatedness and identity in our U.S. context, including for Native Americans. Blood as symbol of inheritance and blood as mechanism of biological inheritance get conflated across cultures, notwithstanding scientific findings that indicate that blood is largely a *carrier* of the mechanisms of biological inheritance (chromosomes and mtDNA) and not the mechanism itself. The culmination of two decades of empirical observations of "white, urban, middle class informants,"[61] Schneider's insights on U.S. conceptions of "blood relatives" are much referred to in kinship literature. They have also been further complicated by newer research on diverse kinship meanings held by Americans of other ethnicities, races, and classes, and by research on kinship after the advent of new reproductive technologies.

In his informants' descriptions of kinship conceptions, articulated through symbolic blood and biogenetics, Schneider demonstrates that his informants move seamlessly across the two ideas as though they are synonyms for the same idea. Scientific discoveries about biogenetic relationships get incorporated into kinship, he explains, as if those new ways of understanding relatedness had always been so.[62] But as new scientific knowledge enters the picture, older meanings do not simply fall away. Sociologist Dorothy Nelkin reminds us that in popular American discourse, "blood often stands in for genes."[63] Schneider shows that understandings of the genetic factors of inheritance get combined with older meanings of blood—with blood continuing to stand in for ideas of inheritance despite an awareness of chromosomal recombination. For example, "The definition of a relative as someone related by blood or marriage is quite explicit in American culture. People speak of it in just those terms, and do so readily when asked. The conception of a child occurs during an act of sexual intercourse, at which time one-half of the biogenetic substance of which the child is formed is contributed by the father, its genitor, and one-half by the mother, its genetrix. The blood relationship is thus a relationship of substance, of shared biogenetic material."[64] Sharing "biogenetic substance" makes "blood relatives," Schneider writes.[65] He explains that for Americans the blood relationship is "real" and "true," never to be truly separated despite legal agreements to the contrary. "Common biological constitution" is seen as producing physical and temperamental commonalities. That a relationship wrought in blood cannot be altered or ended takes on "mystical" qualities. Shared blood constitutes identity, and a loss or denial of the blood relationship is often read as a loss or denial of one's own identity.[66] Blood is identity.

Throughout his richly descriptive ethnography, in which he follows closely the expressions of his subjects, Schneider uses the terms *biogenetics* and *blood* with about the same frequency. He describes the meaning of dividing and tracing blood in a way that also seems to be informed by the concept of genetic recombination: "Because blood is a 'thing' and because it is subdivided with each reproductive step away from a given ancestor, the precise degree to which two persons share common heredity can be calculated, and 'distance' can thus be stated in specific quantitative terms."[67]

But it is not only Schneider's white, urban, middle-class, mid-twentieth-century Americans who interchange DNA and blood as they speak of inheritance. As genetic discourses come to the fore in our society, we see Native American studies scholars import genetic concepts into their analyses of pre-genetic-science eras, when their subjects could not have thought in molecular terms. That is not to say that historical subjects did not sometimes have a grasp on mechanisms for cultivating certain heritable traits in both humans and nonhumans. But they would not have understood such mechanisms as targeting the "molecular."

A key example of a blood-gene conflation is Ward Churchill's denouncement of "untraditional" Indian blood-quantum policies. He writes that by 1830, Native people east of the Mississippi were "genetically admixed," not only with each other but with blacks and whites, too.[68] Churchill finds little evidence that "indigenous societies viewed this increasing admixture as untoward or peculiar." His point is that Indians accepted "admixture" at that time, and therefore it is untraditional to not accept it now. Leaving aside his unexamined assumption that what is "traditional" is stuck at circa 1830, Churchill's analysis gets even stranger. He critiques Indians and their contemporary blood-quantum policies for failing to adhere to a "tradition" that he describes as constituted around the concept of genetic admixture. Yet "genetics" and "admixture" had not yet been articulated in 1830! Of course, Churchill probably means that the Indians of 1830 did not possess those ideas of blood purity and mixture and other nineteenth-century race concepts that circulated throughout both lay and scientific quarters in the United States and Europe. But given his convoluted conflation of the knowledges of different eras, how are we to trust that he has a handle on what the Indians of 1830 actually thought, or what Indians today really think, about heritability, race, and tribal citizenship and all of their complex entanglements?

Churchill travels even further back in time and attributes genetic understandings of race and heritability to Moravian missionaries. He describes them in the late 1700s as preferring to focus their civilizing activities on "mixed-bloods" and as having "the mystical notion that 'Aryan genetics' correlated to . . . intellect and moral capacity."[69] In this instance, Churchill finds fault with missionaries for failing to adhere to genetic knowledge claims (and in their most antigenetically determinist form) that had not actually been articulated in the era in which they lived. Of course, his critique is that the missionaries were among the early racists who saw different human lineages as having different capacities for intellectual and moral development. Fair enough. But Churchill's conflation of historical eras and knowledges also leads him to describe the missionaries as "mystical," again to highlight their antiscientific stance. Of course, their racism may have been quite compatible with the state of race science in that era. Indeed, the missionaries' views as Churchill describes them may also be compatible with some of the *racialist* science being done even today.[70] To clarify, to be "racialist" is to define race as being at least in part biological and not (only) "socially constructed." More racist forms of racialism order supposedly biologically based races into hierarchies, but not all racialism does this. Some genome scientists today are, in fact, avowedly antiracist racialists. They believe that race is in part genetic, but they are explicitly opposed to discrimination based on race.

Underlying both examples is Churchill's assumption that contemporary genetic articulations of heritability and race in their most antiracialist forms are timeless truths that the good guys ("traditional Indians") understand, and the bad guys (missionaries and brainwashed, colonized Indians) do not. Where Churchill attempts to clarify history, he ends up with an ahistorical analysis.

A second example of the conflation of blood and gene concepts comes from a recent treatment of White Earth Anishinaabeg citizenship, blood rules, and cultural values throughout the twentieth century. Jill Doerfler had the good fortune to analyze historical papers produced from 1896 to 1938 that documented how "full-" and "mixed-bloodedness" were understood among early twentieth-century Anishinaabeg on the White Earth Reservation in Minnesota as government investigators attempted to define those terms for the purposes of disbursing treaty annuities. Doerfler shows convincingly that "the concept of blood as a metaphor for racial ancestry and the accompanying terms 'mixed-blood' and 'full-blood' were odd, illogical concepts to Anishinaabeg of White Earth." The archive she examined reveals that concepts of blood were either absent from or indeterminate in Anishinaabeg assessments of who was "mixed." To the contrary, assessments of mixedness, especially as reflected in the indigenous language, had to do with lifestyle rather than biogenetic inheritance.[71]

One of Doerfler's interviewees attributed full-bloodedness to a "way of living" and not to "blood." Doerfler, however, describes his response as rejecting "genetic ancestry" in his definition of full-bloodedness.[72] Writing from the first decade of the twenty-first century, she conflates blood with genes. From our view within a genomic era, her conflation can be seen as a minor semantic imprecision. But it is impossible that Doerfler's subject, unlike Schneider's 1960s white, middle-class Chicago informants, meant blood as a *biogenetic* substance connoting inheritance even as he discounted its importance. Indeed, Doerfler's evidence makes it clear that his thoughts and those of other early twentieth-century Anishinaabeg about blood cannot be conflated with biogenetic ideas! The DNA double helix was not articulated as a form by scientists until 1953 and has since radically altered our language of inheritance.

It is important to note the slip in terminology between blood and genes, because it distracts the reader from the close attention Doerfler pays to early twentieth-century Anishinaabeg thinking on the subject of blood before exposure to dominant U.S. race formations. Peter Wade, in his *Race, Nature, Culture* (2002), lamented the frequent lack of detailed ethnographic data about the actual (biogenetic) properties and outcomes attributed to blood and genes by social actors.[73] Do social actors (especially in the pre-genomic era) intend literal biogenetic meanings when they use "blood"? Is "blood" ever simply biological in its use? Doerfler's rich historical data shows that it is not.

Similarly, Alexandra Harmon, in historical analyses of early twentieth-century Northwest Coast Indian enrollment commissions, calls for more local research on tribal membership histories. She suggests that we are in need of a more solid historical foundation in order to make more informed judgments about the influence of dominant "racial ideology" and U.S. law on the composition of contemporary tribes. Absent a stronger historical foundation, she says it is "conjecture" to claim that tribes have been duped or forced into adopting blood-quantum practices and Euro-American "race codes."[74] Harmon points out that "in the enrollment councils, federal agents did not brainwash or impose their will on Indians; neither did Indians resolve to draw an economically strategic, racially defined boundary around themselves. Rather, officials and Indians participated in a prolonged discourse that I would characterize as incomplete mutual education and accommodation."[75] Harmon refers to the commissions as "an unprecedented conversation—one that would take place in many tribal communities and continue for decades—about what it meant to be Indian in the twentieth-century United States."[76]

Unlike Doerfler's or Harmon's scholarship, Churchill's does not cite rich data to back up his charge that the missionaries believed that mixed-blooded Cherokees had greater civilizing potential due to the material properties in their blood, although it is possible given both the state of early race science at the time and ensuing colonial policies. Not caring much to defend the missionaries and the complexity of their sixteenth-century race thinking against Churchill's charges, Churchill missteps (and Doerfler does, too) in glossing historical blood meanings for different actors as "genetic" or "not genetic." Molecular articulations did not come into the matter for anyone. To imply that they did for some (even if the goal is to highlight from our viewpoint the more liberated thinking of indigenes) is to attribute universality and timelessness to gene concepts. Again, this move undermines the broader project of defetishizing genetics in our concepts of full personhood.

Churchill repeatedly exchanges "genes" for "blood" in his critique of tribal enrollment policies. In one instance, he cites Joseph Greenberg's classic and widely referenced linguistic grouping of American indigenes,[77] in which three language groups correlate "rather well," there being only "three, or perhaps four, discernible gene stocks" across the two continents.[78] Churchill's point is to show the lack of genetic distinction between the hundreds of Native American tribes, an assertion with merit. But in using genetic science as an absolute standard of truth even in our genomic era, Churchill not only fetishizes genes, he also ignores any complexity in tribal blood talk and politics: Indians are both scientifically ignorant and self-colonizing.

Of course, the jump that Churchill makes to tribal blood criteria as *genetically unscientific* is a straw-man argument. Tribes do not defend blood-quantum rules as biological science. They may not be explicit about the semiotic nature of the blood they invoke, and many today may indeed invoke blood as a symbol for their basic understandings of molecular mechanisms for the inheritance of traits. But again, this is a contemporary manifestation, as Doerfler's work shows, as our thinking about inheritance has been widely figured as "genetic" only since 1953. Tribal blood rules have been in effect and evolving for much longer. Indeed, we are only beginning to learn tribal members' complex and evolving thoughts on blood meanings with the work of younger scholars. What we do know is that blood-quantum determinations have been based upon fractions enumerated on paper by bureaucratic agents of both the state and tribes and have sometimes been rooted in interviews and other empirical observations by late nineteenth- and early twentieth-century state agents.[79] Blood as a potentially real material phenomenon is not terribly relevant to identity. Blood fractions are not enumerated based on examinations with laboratory prostheses of blood the fluid or of the genetic material found in cells. Blood quantum is a materialist practice only to the extent that it involves paperwork. Our societal discourse today moves so easily between semiotic and material meanings of blood and genes that we seem to forget that. As I write this, I read yet another critique of blood quantum placed prominently in the *New York Times* in which Native American literature scholar and fiction writer David Treuer conflates blood and genetics: "Blood quantum . . . has always been about exclusion. . . . Who is and isn't an Indian is a complicated question, but there are many ways to answer it beyond genetics alone."[80] I, too, was once guilty of closing the distance too much between blood and DNA.[81] But over years of studying this issue, I have learned that we need to keep clear the difference between biogenetic properties and blood quantum as a semiotic and bureaucratic object constituted through other forms of science, namely, the *social* and *policy* sciences.

Given its currency as a biologically deterministic fact in contemporary discourse and given the measurability of the material of DNA (although precisely what is measured and what that means are in question), will Indian blood cede semiotic ground to Native American DNA in the constitution of key categories? "Native American," "Amerindian," "tribe," "Pequot," "Ojibway," "Nuu-chah-nulth," "indigenous," "isolates of historic interest," and the like can be sliced in multiple ways depending on who uses such terms, in which institutional and historical contexts, and what is at stake for whom. How do "Indian blood" and "Native American DNA" differently condition categories of race, population, lineage, tribe and First Nation? Which discursive practices will prevail in which contexts? Before grappling with those questions, we need to understand the emergence of the American Indian "tribe" and tribal "members" or "citizens" as a social and political phenomenon in the United States since the late nineteenth century. And we need to understand how the formation of an "American Indian" or "Native American" race category informs the constitution of the tribal body politic. Blood has supported the rise of both "tribe" and "race," categories that overlap but are not synonymous. With that understanding, we can begin to imagine the opportunities and the barriers for DNA to (re)configure those categories in the twenty-first century.

Blood-Quantum History, Critiques, and Defenses

Scholars typically analyze blood meanings in contemporary tribal citizenship as a leftover colonial imposition that can be traced back to outmoded Euro-American conceptions of blood and race that informed the construction of tribal rolls in the late nineteenth and early twentieth centuries.[82] Scholars generally explain that "blood quantum," as it has come to be called, emerged as an incisive social technique for managing Native American lands and peoples. U.S. federal agents settled upon blood as a mechanism to break up collectively held Native American land bases.[83] Over the last century, critics argue, many Native American tribes and others have been brainwashed by such biologically essentialist thinking. First, blood rules were imposed on them; now, tribes are self-colonizing. Today, despite social-constructionist analyses of race to the contrary and the logic of tribal political sovereignty, U.S. tribes are loath to forgo blood notions in citizenship. They foolishly continue to believe that blood (alone) matters. This is a standard characterization.

For those who are unfamiliar with the controversial history of blood quantum, a brief overview is in order. But I will complicate that history beyond the more standard, top-down characterization of tribes as simply intellectually colonized by old-fashioned race and blood concepts. Indian and tribal identification and the constitution of the first or "base rolls" of approved tribal members can be traced to the requirements of the General Allotment Act of 1887, also known as the Dawes Act, after Massachusetts senator Henry M. Dawes, who sponsored it.[84] The Dawes Act divided communally owned reservation lands into individual 160-acre, 80-acre, and 40-acre allotments. But before commonly held Native American lands could be distributed, lists of "tribe members" had to be constructed. Although the Dawes Act specified no criteria by which formal lists of Indians should be compiled, it noted that land should be allotted according to "belonging" or to "tribal relations." After rolls of Indians were determined, parcels of land were then allotted to individual Indians depending on their status respectively as a head of family, as an unmarried adult, or as a minor. Indians deemed to be "half-blood" or less were given full title and, with it, U.S. citizenship. Indians who were deemed to be more than half-blood had title held for them in trust for twenty-five years. The Dawes Act was built on the common assumption that individual land ownership would assimilate Indians. It was hoped that it would make them individualistic farmers and better subjects for a capitalist economy. On the other side of the nurture/nature coin was the idea that those with less Indian and more European blood were more advanced on the evolutionary road to civilization. These "mixed-bloods," therefore, had greater autonomy in land tenure.[85] They could legally sell their land to others. Indeed, in addition to promoting the assimilation of Indians into dominant U.S. culture, the act worked even more effectively to assimilate Native lands into the land base of the still-developing (white) nation. After distribution of Native American allotments to those deemed eligible, the extra land was sold off or given to white settlers. Therefore, tying blood quantum to land tenure aided the project of dispossessing Indians from their land.

The standard story of blood critics describes federal agents charged to implement allotment at individual reservations as having significant autonomy in figuring out to whom land should be allotted and with what title. Thus, dominant race thinking of nineteenth- and early twentieth-century whites predominated in the allotment process.[86] But a closer look at the historical archive shows that in the early days of constructing tribal rolls, federal agents sometimes sought Indian input on who should be a tribe member.[87] In addition, Native American and indigenous studies scholar Joanne Barker shows that racial ideas were only part of what motivated federal agents. Expediency was another factor. Agents were being pressured to quickly complete the allotment process in order to expedite statehood in what had been Indian territories. They required a convenient mechanism to determine which Indians qualified for which type of allotments. Blood quantum was that bureaucratically efficient means.[88] So, although older forms of race thinking inform evolving tribal blood policies, that is only part of the story.

Self-Determination and Tribal Citizenship: "Tribal Blood" Displaces "Racial Blood"

Since the turn of the twentieth century, tribal blood rules in the United States have become more complicated and varied as the federal-tribal relationship has become less patronizing, as tribes' governance authorities have increased, and as tribes have responded to political and economic pressures and demographic shifts. In fact, despite outspoken critiques against tribal blood rules as self-imposed elimination (some critics ask, with minimum blood-quantum rules, how long before there are no Indians left to enroll?), legal scholar Kirsty Gover shows that in the last several decades, tribes have increasingly moved away from the racial mechanism of total "Indian blood" to a genealogical mechanism of "tribe-specific blood." They eschew the monolithic racial category of "Indian," but keep "blood" in order to "maintain and repair continuity during shifts in federal Indian policy and tribal demography."[89] Gover shows that in tribal constitutions dating from the 1930s, parental enrollment coupled with on-reservation residency rules dominated. But World War II military service and economic opportunities, including jobs in cities, as well as federal policy after World War II (termination and relocation) prompted much higher rates of exogenous marriage (marriage outside the tribe).

It became difficult for younger generations to meet blood-quantum requirements. Major shifts in reservation demographics, in turn, prompted changes in tribal membership rules. Five percent of Indians in 1940 were urban dwellers, but by 1970, one-half lived in cities. As reservation continuity was disrupted for many, parental enrollment and residency rules were not sustainable. Tribal-specific blood rules attempted to repair that discontinuity (children of urban Indians could be enrolled). In addition, lineal descent coupled with tribal descent imposed limits on the number of potential enrollees, especially after the 1970s and the federal Indian self-determination policies, when tribes became wealthier.[90]

"Self-determination" policies begun in the 1960s and 1970s repealed earlier federal policies of tribal termination and some of the most patronizing federal government oversight. Self-determination has devolved to tribes' rights to manage land and natural resources, education, housing, and health-care programs. Self-determination also devolved on most tribes the right to determine and manage their own membership or citizenship criteria, a right now seen as a key to the exercise of tribal sovereignty, as well as a right rooted in tradition.[91] Yet the federal government still "urges limitations on descent rules as a condition of tribal sovereign status," especially for "newly recognized or restored tribes." (For existing tribes, federal oversight is lighter.) As Gover explains, the Department of the Interior has "a strong interest in the composition of tribal base rolls" and has "threatened disapproval" of draft constitutions for "over-inclusivity of the base roll."[92] For the feds, there must be a historical—read biological, read blood—continuity between Indians of old and today's tribal membership. For tribes, too, blood rules continue to have appeal. Indeed, their importance has grown. But we should be wary of assuming that tribal and federal understandings of blood and reasons for instituting blood rules are in sync. Gover finds that federally recognized U.S. tribes became more likely to institute blood rules after 1970 and self-determination than to rely on parental enrollment and reservation residency rules that were dominant before the mid-twentieth century, when most Indians lived on reservations.[93]

The Cherokee Nation, for example, with a large and far-flung diasporic population, "grants citizenship only to lineal descendants of Cherokees by blood," that is, those who trace their ancestry to the Cherokee blood rolls.[94] Unlike in many other tribes, Cherokee Nation enrollment applicants need not meet a certain blood-*quantum* requirement. But a lineal blood connection, however thin, matters very much, as is demonstrated by recent controversies involving the disenrollment of the descendants of Cherokee freedmen, former slaves naturalized as Cherokee during the nineteenth century.[95] Despite Ward Churchill's assertions to the contrary,[96] lineal biological descent is not simply a more "traditional" alternative to the colonial mechanism of blood quantum. In addition to being a pragmatic response to demographic shifts, David Schneider's ethnography demonstrates that, like blood degree, lineal descent is also no less a Euro-American analytic.[97] It is notable that Native American tribes since World War II have shown themselves unwilling to dispense with blood criteria completely. They may lower or dispense with a blood-*quantum* requirement, but a blood *link* is still imperative. Continuing attachments to blood concepts, despite tribes' movement away from the federal race mechanism of total Indian blood, are also demonstrated in their increasing refusals since World War II to enroll spouses and adopted offspring who are either non-Native or descended from other Native American tribes.

Anthropologist Beatrice Medicine and Native American studies scholar Elizabeth Cook-Lynn have also argued for the pre- or extra-colonial importance of certain blood concepts in constituting the Native American tribe.[98] As David Schneider saw with his white, urban, mid-century, middle-class informants in *American Kinship*, Medicine sees the blood relationship as key in how Native America tribes reckon kin. But although nuclear family and direct lineage relations are included in the idea of "tribe," Medicine sees the important blood relationships as broader than that. Her work focuses on identity that emerges from one's links to multiple tribally affiliated individuals. She sees the identity of a person of Indian descent as tied to tribalness, that is, to a social grouping based upon biological relationships. For Medicine, the blood-infused but more-than-biological relationship of an individual to her tribe is fundamental to her Indianness. However, Medicine also notes that for her informants, tribal rootedness can override low degrees of Indian blood. Any blood at all, *if* combined with a more-than-biological relationship, indeed mattered more than blood alone and its symbolic measurement in the form of blood quantum.[99] In short, the tribe is a networked set of social and cultural relations based on biological relatedness.

Elizabeth Cook-Lynn, who also appreciates cultural bases for indigenous identity,[100] still relentlessly focuses on tribal political sovereignty as the source from which tribal citizenship stems. And unlike some other scholars, she does not translate that appreciation for a politically mediated tribal identity into a call for non-descent-related cultural criteria. She wants tribes to be recognized first and foremost as legal entities, not simply cultural entities.[101] And she does not advocate blood quantum as a must for conferring tribal citizenship. But, like Medicine, Cook-Lynn acknowledges blood concepts (whatever those may be) as long-standing ideas structuring who belongs. Being Indian or a member of a particular indigenous nation is precisely about citizenship, but it is citizenship based in part on blood, which has "been a tenet of survival and identity in native enclaves from the beginning and continue[s] to be."[102] Furthermore, Cook-Lynn sees upholding tribes' rights to confer citizenship, even if they do it in ways that displease some, as critical to perpetuating indigenous legal authorities and land claims that guarantee indigenous political survival.

Many in Native America also worry about idealized American longings for romanticized Indianness that sometimes get coupled with increased economic benefits in an era of self-determination. Tribal communities sometimes feel they are under assault by people with tenuous or nonexistent connections to their communities yet who want access to cultural knowledge or to cultural sites for personal identity exploration and sometimes for profit. Such psychic and material longings have prompted some to seek a tribal identity.[103] As some critics point out, some tribal councils make revisions to enrollment criteria in support of short-term political and economic goals, such as excluding certain disagreeable families, political opponents, or even previously qualified tribal applicants. And there is no doubt that antiblack racism informs the resistance of some within the Cherokee Nation to keeping freedmen's descendants on the rolls. However, across Indian Country, there are also legitimate worries about real, long-term material and economic risks to tribal communities and about the overwhelming resources needed to maintain tribal government infrastructures, including health care, housing, and education—worries that prompt tribes to continue emphasizing biological connection to the tribe in particular ways. This is why pleas or admonitions for a more "decolonized" tribal enrollment based on social- and cultural-competence criteria (such as doing community service on reservations or in historic homelands; knowing the tribal history, culture, and politics; knowing the tribal language; taking an oath of allegiance to the tribal nation; and proving that one is of "good character according to the tribe's traditional code of morality"),[104] used alone or in conjunction with more liberal blood rules, either fall on deaf ears or are seen as overly idealistic. Monitoring biological connections is bureaucratically difficult and costly enough; the fear is that floodgates would fly open with the possibility of cultural conversion.

More Than Biological Essentialism: Blood Quantum Tallies Relatives

Like Medicine and Cook-Lynn, historian Melissa Meyer complicates a too-easy biology-culture binary that blood critics see as structuring tribal-enrollment debates and options. Both Cook-Lynn and Meyer analyze tribal-enrollment criteria through a double lens of tribal political sovereignty coupled with the concept of culture, with blood relatedness acting as a proxy for culture. Meyer points out that, rather than tribes dumbly thinking that the material substance of blood equals culture, today they are using tribal-citizenship criteria involving blood that in fact strive "to reflect some sort of valid cultural affiliation." She credits tribes with understanding that blood is no guarantor of cultural affiliation and that some people with "legitimate cultural ties will be eliminated."[105] But tribes assume that higher blood degree will increase the odds of true affiliation. Indeed, in order to meet tribal blood-quantum standards, tribally affiliated individuals must make up a sufficient proportion of one's genealogically documented progenitors. For example, if a tribe's quantum rule is one-quarter blood, this means an applicant must be able to document one "full-blood" grandparent, two "half-blood" grandparents, or the like). More, rather than fewer, tribally affiliated relations are seen as raising the probability that a tribal member will actually be culturally affiliated, compared to one who simply fulfills a lineal-descent requirement.[106]

For readers who are familiar with my earlier work,[107] my thinking on these matters has evolved, in particular regarding the distinction to be made between "race" and "tribe" and the way these categories are informed by different blood rules. In a 2008 chapter, "Native-American-DNA.coms," I described the move toward lineal descent in tribal enrollment as compatible with the *racial* logic that supports genetic-ancestry testing.[108] I explained, "The technique of blood quantum accounts for but goes beyond lineal descent to focus on documented blood links between an individual and the *group* via blood links to a multiplicity of documented tribal individuals."[109] But Gover's richly empirical work has complicated my view of the relationship between lineal descent and blood-quantum rules. Sometimes the two are used together to document one's relatedness to a tribe in a way that also opposes the racial mechanism of the requirement for total Indian blood promoted for so long by the federal government in its Native American identity policies. On the one hand, Gover explains that "the federal government's primary concern is to ensure that tribes are comprised of Indians and whether those Indians have an ancestral tie to the 'accepting tribe' is of secondary importance." On the other hand, tribal blood rules require ancestry within the tribe, and sometimes they consider other Native American ancestries, but that is of secondary importance.[110] Yet, when used alone, as I have noted elsewhere, lineal descent from a single ancestor on a base roll can be seen to be compatible with Euro-American race logic.

I offer a personal example—my own blood-quantum designations—to illustrate precisely Gover's analysis of blood rules, as well as to demonstrate the considerable intertribal social mixing that happened throughout the twentieth century as a result of demographic shifts. On the reverse of my tribal identity card issued by the Sisseton-Wahpeton Oyate (SWO) of the Lake Traverse Reservation in South Dakota, my tribal ancestries are classified into the following complex (for the uninitiated) blood-quantum categories:

- Degree of SWO Indian Blood: 1/32 (my maternal great-grandfather was classified as ¼ Sisseton-Wahpeton).
- Other Sioux Blood: 1/16 Flandreau Santee (my maternal great-grandfather was also classified as ½ Flandreau Santee).
- Other Indian Blood: 1/16 Turtle Mountain Chippewa (my maternal great-grandmother was classified as ½ Chippewa); and ¼ Cheyenne and Arapaho (my maternal grandfather was classified as "full-blood" Cheyenne and Arapaho).

In total, I am documented as having "13/32 Total Indian blood." SWO enrolls applicants based on total Indian blood of at least ¼, with a genealogical trace to the Sisseton-Wahpeton Oyate. Thus, although I am documented as only 1/32 Sisseton-Wahpeton, I meet the ¼ minimum *Indian* blood, and I am in. But according to Gover's analysis, it gets a little more complicated than that. I am one of those individuals with "high aggregate multi-tribal blood quantum" when enrolled according to total "Indian blood." But that Indian blood is referenced in tandem with the concept of lineal descent. In this case, as Gover notes, "Indian blood serves to qualify a tribe-specific descent rule."[111] Again, this particular blood rule is not simply about enrolling an Indian. It is also about counting particular ancestors. It is also a genealogical measure.

A stricter form of "genealogic tribalism" is the "tribal" blood-quantum rule. To illustrate, I was previously enrolled in my maternal grandfather's tribe, the Cheyenne and Arapaho Tribes of Oklahoma (C&A). I gave up my affiliation with C&A (neither tribe allows dual enrollment) in order to enroll in SWO.[112] Having been raised in South Dakota and Minnesota among my Dakota relatives, I feel more culturally Dakota than Cheyenne and Arapaho. I have never lived in Oklahoma; I simply tried to make sociality and culture better match my tribal-nation citizenship. C&A at present does not consider other tribal lineages or total "Indian blood" in granting citizenship. Instead, it requires a minimum of ¼ Cheyenne-Arapaho blood or ¼ "tribal blood." Interestingly, although I feel I can make stronger social claims to being Dakota, I have a higher blood quantum and meet the stricter requirements of the Cheyenne and Arapaho Tribes. These rules and fractions seem arcane, even shocking, to nontribal people in the United States, but we in Indian Country are intimately familiar with such practices even as they differ from tribe to tribe. Many of us have ancestors and relatives in multiple tribes.

In summary, the story of tribal citizenship in the twentieth and twenty-first centuries is one in which dominant cultural notions of race—federal "Indian blood"—have pushed and been pushed against by tribal peoples' own ideas of belonging and citizenship. The "tribal blood" fractions can be seen to represent a counting of tribal relatives in the conferral of membership. That is not to say that many of us, native and nonnative alike, might not believe vaguely that the concepts of Indian blood or tribal blood represent underlying biological properties—that we are not sometimes being biologically deterministic in our use of these concepts. But the counting of relatives and establishing a genealogical connection to them is also clearly at play in our blood talk. We use the language of blood and blood fractions while keeping in mind a specific world of policy *and* while bearing in mind that that language is shorthand for what we know is a far more complicated story of our lineages. When I cite those fractions, I think of my grandparents and great-grandparents. I remember their names and their parents' and grandparents' names. I remember how, through both dispossession and restricted choices, they came to be on the particular reservations now denoted in my blood-quantum fractions. These are relatives whose stories have been passed down to me, sometimes from their own mouths. I am not alone in Indian Country in this practice of accounting.

Nature-culture binaries and social-science refutations of them are so influential that not only is it intellectually confusing but also it feels academically countercultural to work one's way through to a more nuanced understanding of how nature and culture actually relate rather than mutually exclude each other. I don't suggest—and I don't see Medicine, Meyer, or Cook-Lynn as suggesting—that cultural knowledge inheres in the physiological properties of blood, a notion that blood-rule critics would rightly characterize as biologically essentialist. The relationship that these scholars see between blood and a people, and which they see tribes as seeing, is more complicated than that. Remember Doerfler's analysis of early twentieth-century Anishinaabeg understandings of blood? They meant not-biology, or something more than biology. Although contemporary tribal peoples may be more influenced by mainstream blood and gene talk than those Anishinaabeg would be, there is more going on with tribal understandings of blood than simply essentialism. Both lineal-descent and blood-quantum concepts invoke blood semiotically to organize Native American identity differently but, as Gover shows, in potentially complementary ways. Blood rules can and do have roots in the Euro-American racial thinking of earlier centuries. But "blood" may also retain symbolic meanings rooted in indigenous thought, and blood rules require the counting of one's tribal relations and ancestors. Indeed, as Gover shows, the move by tribes away from total Indian blood to tribe-specific blood, or the coupling of blood quantum with lineal descent in a specific tribe, is a move away from that earlier race thinking and toward a new "genealogic tribalism."[113] This has similarities to J. Kēhaulani Kauanui's argument against the State of Hawaii's definition of "native Hawaiian" based on a 50 percent blood-quantum rule. Instead, Kauanui argues for a more traditional mode of genealogical accounting in which Kanaka Maoli—a term literally meaning "real or true people" but used to reference Hawaiian indigenes[114]—trace their genealogies *up and down* both matrilineal- and patrilineal-descent lines *and across* the generations. They include biological and social relations and they have several modes for incorporating nonbiological relations into the family unit and their genealogies.[115] Kauanui emphasizes the inclusivity of Kanaka Maoli kinship. She argues against blood quantum, but she also shows that blood is not inconsequential in Kanaka Maoli kinship. It is just not the only way to have kin. Kin is also made socially. Alas, we, too, retain ways for making kin socially and ceremonially in Indian Country, but making our citizenship rules account for these practices will be a politically and economically difficult task.[116]

The work of these scholars adds more nuance and texture to our understandings of the role of blood in tribal citizenship and identity. Gover's empirical evidence and her analysis of how blood rules and lineal descent are used in tandem in enrollment today complicate our understanding of the nature of blood rules in tribal citizenship. Yet I must still emphasize the difference between lineal descent and blood quantum in order to provide context for my claim that DNA rules are not simply newer forms of blood rules, nor are "blood" and "DNA" as material-semiotic objects interchangeable ideas. The concept of linear genealogical descent also conditions the uses of genetic-ancestry testing by the American public in ways that are beginning to touch indigenous citizenship processes. The DNA genealogies that are documented by ancestry tests (again, different from the DNA parentage tests that tribes use) and that are co-constituted with hegemonic U.S. race concepts are not yet compatible with the particular biological relationships that tribes privilege. Yet enrollment staff from several tribes told participants at a 2010 national tribal enrollment conference that they had received enrollment applications with commercially purchased genetic-ancestry test results included.[117] This happens even though federally recognized tribes do not accept genetic-ancestry results as appropriate documentation for enrollment and do not advise applicants to submit such documentation. In fact, enrollment officers express confusion about the science behind genetic-ancestry tests and seem to have little familiarity with them.

Whereas other scholars demonstrate that biology and culture have a more complex relationship in indigenous identity-organization and citizenship processes than simply a mutually exclusive dichotomous relationship, this book argues against the conflation of blood and gene concepts on the biology side of that unhelpful binary. The pressures of a colonial state certainly continue to inform indigenous citizenship and identity formation, but Native American tribes have not been passive in these processes. They have also been defensive and sometimes generative in their regulatory responses, if not completely throwing off the yoke of colonialism—a tall order indeed.

2

THE DNA DOT-COM

Selling Ancestry

> The gene is the unit of Life. The soul is the unit of Humanity. We know the alphabet of Life, we have unravelled the code. But remember, like words, DNA has significance beyond the sum of its parts.
>
> —David Bromfield, DNAPrint.com

In the early 1960s, researchers began applying new genetic techniques to traditional anthropological questions.[1] The new science was coined "molecular anthropology." Today, researchers around the world use a growing arsenal of techniques to study ancient human migrations and the biological and cultural relationships between human groups in different geographic locations. Researchers draw blood and in other ways capture DNA from human bodies, both living and dead, the world over. "Native American," or "Amerindian," populations and bodies are popular objects of study. Researchers search in specific locations of an individual's genome for certain molecular sequences, those "genetic signatures" of ancient peoples and the sources of today's human populations.

Since 2000, so-called Native American DNA (and "Indo-European," "sub-Saharan African," and "East Asian DNA," to name a few others) is detectable through a self-administered cheek swab.[2] The desire for knowledge of particular types of molecular sequences in one's genome has fed a commercial phenomenon, what has come to be called direct-to-consumer (DTC) genetic testing. By 2006, 460,000 people had purchased such tests,[3] and the technologies continue to capture the imaginations of especially American and British consumers.[4] Today, approximately thirty companies market genetic-ancestry tests to the public.[5] Back in 2007, the prominent "genetic genealogist" Blaine Bettinger predicted on his blog, *The Genetic Genealogist*, that "the 1 millionth genetic genealogy customer will push the 'buy' button as early as 2009." Bettinger scoured not only academic papers but also company Web sites for disclosures of numbers of tests sold. Based on what appeared to be a growth rate of "as much as 80,000 to 100,000 per year," that number may have reached as high as 1.3 million at the end of 2011.[6] As will become obvious throughout this chapter, I tend to trust the data produced by skilled genetic genealogists, although I may sometimes disagree with their interpretations.

Who buys genetic-ancestry tests? A variety of individuals with different historical and personal needs purchase the tests. A key part of the market is genealogists, or "family-tree researchers," who may use a genetic-ancestry test to search for where and to whom they might trace their immigrant, African, or Native American roots. In which contemporary villages, ethnic groups, or tribes can they find genetic ancestry in common? These genetic genealogists look specifically at Y-chromosome lineages in order to help with their surname-project research. They are interested in finding out whether a man named Smith in Westport, Connecticut, for example, is related to another named Smith in Blackpool, England. Do those two males share Y-chromosome markers that can link them definitively to a long-ago common ancestor, and then to each other? I address the work of genetic genealogists, an influential body of researchers-cum-consumers, more fully in the next chapter.

Some consumers combine genealogy research with more explicit political and economic desires. Former citizens of Native American tribes, ejected for reasons having to do in part with the financial stakes of membership, have used DNA in their attempts to prove their biological links to tribes.[7] There is a lot at stake for such individuals: not only health, financial, and educational benefits but also one's deeply held identity. Sheer financial opportunism can also be a motivator for DNA testing. A news article posted to a genetic-testing company Web site notes that one customer used its test during the trial phase to "prove he qualified for a business venture exclusive to Native Americans."[8] Anecdotal evidence also suggests that applicants to Ivy League and other top-ranked schools who want an affirmative-action leg up in competitive admissions processes have used DNA tests to back up their personal decisions to self-identify as racial or ethnic minorities.[9]

In this chapter, I analyze the technical and cultural production of six DNA-testing companies: DNAPrint Genomics, DNAToday, Genelex,

GeneTree, Niagen, and Orchid Cellmark. I examine their work in relation to two categories of conceptual and social organization, "race" and "tribe," that are fundamental to how Native American history and identity gets understood and regulated. In chapter 1, I treated the histories that inform how those terms are defined, where they overlap, and where they diverge in the American scientific and popular imagination. Here, I examine how historically and culturally contingent notions of "race" and "tribe" inform the marketing and interpretation of company technologies, and to what effect. How are genetic-ancestry tests partly constituted by such categories, and how does the work that consumers make tests do loop back to reconfigure and solidify "race" and "tribe" as genetic categories? I look at how genetic technologies and societal notions of race and tribe are co-constituted in the early twenty-first century.

Five of the six companies I examine have prominently targeted the Native American–ancestry market: DNAToday, Genelex, GeneTree, Niagen, and Orchid Cellmark. They especially foreground the concept of tribe in their marketing. Two of the five companies, DNAToday and Orchid Cellmark, have marketed the DNA profile (sometimes also referred to as a "DNA fingerprint") or common paternity test for use in tribal enrollment. This test is a different type of genetic analysis, not the more typical "genetic-ancestry" product. DNAPrint Genomics does not specifically target Native American ancestry, but it has developed a complex, patented autosomal, or cross-genome, DNA test that provides a fascinating case study in how notions of race get (re)configured in complex ways in genetic tests. We live in turbulent economic times. Two of the companies have declared bankruptcy since I began studying them, in 2003: DNAPrint, in 2009, and DNAToday, in 2006. But both remain important cases for understanding how concepts of tribe and race get deployed in relation to Native American identity. DNAPrint's unique technology has been licensed to another company, and other companies continue offering services to Native American tribes that are very similar to that of DNAToday, both technically and in terms of the gene talk they offer up to tribes.

Before I delve into analyses of each of the six companies, the reader should take note that the "technical" culture—meaning both the materiality of molecules and statistical ways of knowing them—and nonscientific cultures are densely entangled in the work of this industry.

Native American DNA as an object of knowledge is best described as neither simply material nor purely symbolic. It is neither discovered nor simply made up. Think of Donna Haraway's description of the more general gene as a "material-semiotic object of knowledge" that is "forged by heterogeneous practices in the furnaces of technoscience."[10] Native American DNA is, likewise, constituted by relations of humans and nonhumans in the practices of technoscience. Thus, there is no understanding to be had or critique to be made of this industry or its objects of knowledge that does not bring the technical/material and the cultural simultaneously into play. Despite that, I sometimes describe DNA-test shortcomings as having more to do with technical issues or more to do with social or (a)historical assumptions on the part of scientists. I also disclaim a deep treatment of tests' technical aspects; others do that better.[11] Yet I cannot fully avoid technical critiques, and those other writers are also deeply attentive to cultural production. Again, the "cultural" and the "technical" cannot be disentangled. If I appear to try to disentangle them for just a moment, it is for the purpose of a readable analysis.

Gene Fetishism: The Myth of the Master Molecule

By describing the gene as a material-semiotic object of knowledge, Donna Haraway gives us a conceptual alternative to simpler forms of social constructionism and to something she calls "gene fetishism." Haraway's explication of gene fetishism informs my analysis of company marketing and spokespersons' claims about the work that Native American DNA can do. Haraway defines "gene fetishism" in relation to Marx's notion of commodity fetishism. Marx saw that in the process of exchange, as objects become commodities, they become transcendent, evolving almost magically. They become signs in and of themselves. Commodities come to be seen as autonomous, objective things, obscuring and displacing the social relations between the humans and, Haraway would add, the nonhumans involved in the production of such objects.[12]

Gene fetishism also obscures complex interactions between humans and nonhumans. Haraway explains that "gene fetishism . . . is about mistaking *heterogeneous* relationality for a fixed, seemingly objective thing."[13] Although genes, the body, and life itself are actually constituted by many complex interactions between humans and nonhumans, gene fetishism leaves us with an impoverished understanding of DNA as a "master molecule" or as the "the code of life." The molecule itself comes to be seen as the source of value[14] and, in the case of Native American DNA, as a source of identity and potentially, then, a foundation for compelling claims. Far more complex political histories of relations and power in the constitution of Native American bodies, citizenries, and life get obscured, which matters. Those historical interactions are critical to a complex understanding of U.S. history. The molecule-made-transcendent is a poor substitute for that historical complexity. When genes or DNA stands for complex interactions between humans and nonhumans, DNA helps to de-emphasize the political histories upon which Native American sovereignty claims get articulated. I believe this obfuscation is ultimately damaging to Native American self-governance claims, as I hope to show by the end of this book. Of course, not all instances of gene fetishism result in adverse effects for the social actors involved.

DNAPrint Genomics and Race: When Mixture Is Predicated on Purity

DNAPrint Genomics declared bankruptcy in March 2009, a victim of the current financial crisis.[15] But its patented and popular AncestrybyDNA™ test was quickly licensed to the DNA Diagnostics Center.[16] AncestrybyDNA™ provides a detailed and highly visible example of how the concept of race gets understood and deployed in the DTC genetic-testing industry. From a technical perspective, AncestrybyDNA™ is unique, because it is an autosomal ancestry DNA test that surveys for "an especially selected panel of Ancestry Informative Markers" (AIMs) across all twenty-three chromosome pairs. Mark Shriver, the former technical adviser to DNAPrint, and his colleagues define AIMs as "genetic loci showing alleles with large frequency differences between populations." Shriver et al. propose that "AIMs can be used to estimate biogeographical ancestry (BGA) at the level of the population, subgroup (e.g., cases and controls) and individual."[17] Because AIMs are found at high frequencies within certain groups and at lower frequencies in others (sometimes very rarely and other times not so rarely), DNAPrint surveyed markers and used a complicated algorithm to estimate an individual's BGA percentages. Most other ancestry DNA tests look at markers on the Y chromosome or on the mtDNA, which are genealogically very informative but more limited in the ancestral lines they survey.

DNAPrint's final version of its test, AncestrybyDNA™ 2.5, surveyed approximately 175 such AIMs, a number that DNAPrint continually worked to increase. (A mere 175 markers is a minute sample of the total number of base pairs in the human genome, between 3 and 4 billion.) The most touted AIM surveyed is the Duffy Null allele. It is found in virtually all sub-Saharan Africans, reaching 100 percent frequency in some of those populations, probably because it confers resistance to malaria caused by *Plasmodium vivax*, a malarial parasite.[18] Research shows the marker to be rare in populations outside of Africa.[19] The assumption is that if an individual has the Duffy Null allele, she or he must have relatively recent ancestry (that is, on the order of several thousand years) in sub-Saharan Africa, an area of Africa where *Plasmodium vivax* is common.

However, the Duffy Null allele, an exemplary AIM discussed at length on DNAPrint's Web site, is not typical of AIMs. It is genetically selected for because it confers an important trait: resistance to a potentially life-threatening parasite. Deborah Bolnick explains that the "vast majority of human genetic variation does not follow this pattern," but AncestrybyDNA™ "emphasizes the very few markers that may do so."[20] The full range of AIMs used to calculate "ancestral proportions" is proprietary information, which makes peer and public review of the science difficult.[21] Only a handful of scholarly articles were ever cited by AncestrybyDNA™ on the company Web site, which collectively disclosed about one dozen such markers.[22]

Genetic genealogists have also expressed concern with the lack of disclosure about the research and methods backing up the tests. Lack of detail on markers surveyed can frustrate their genealogical research. An exchange on January 11–12, 2004, centered on particular markers surveyed by the test. Ann Turner, founder of the RootsWeb Genealogy-DNA-L Listserv, cross-referenced academic papers and genetic-marker databases in order to determine and share with the list several markers used by DNAPrint to calculate Native American and other BGA percentages. Turner cross-referenced markers noted in the "sequences.pdf" file, which DNAPrint provided to customers as scientific background along with test results, with markers noted in table 1 of Shriver et al.'s

2003 paper "Skin Pigmentation, Biogeographical Ancestry, and Admixture Mapping."[23] Turner found several examples, in addition to the Duffy Null allele, of markers that DNAPrint used in calculating Native American and other BGA categories. Consider DNAPrint marker 1141, otherwise known as SGC30055*1. It occurred in 0.753 of Native Americans sampled, just 0.054 of African Americans sampled, and 0.511 of European Americans sampled. It is good for distinguishing Native American from African ancestry but not so good for distinguishing Native American from European ancestry. DNAPrint marker 1116, otherwise known as WI-17163*3, was found in 0.690 of Native Americans sampled but in only 0.175 and 0.054 of European Americans and African Americans sampled, respectively.[24] Such markers are helpful but not nearly as clear-cut as the Duffy Null allele for distinguishing BGA.

Other methodological questions also come to mind. First, of the 175 markers surveyed, what is the proportion that is informative for Native American ancestry? Do the 175 markers fairly evenly cover different ancestry categories, or are there many more that help distinguish African or European BGA? Second, did DNAPrint sample randomly across the genome in choosing markers to survey? We don't have any evidence other than the fact that the company said it did. But we do know that there is actually a limited number of known AIMS.[25] Therefore, we must ask whether there actually *are* known AIMs "across the genome." Furthermore, did DNAPrint use skin-pigmentation markers simply because they're convenient and not necessarily because they were the best markers for DNAPrint's purposes, that is, for determining "unique lineages"?

Bolnick has noted additional technical problems with AncestrybyDNA™ that challenge DNAPrint's claim that the test can detect "precise ancestral proportions" and that challenge its overall claims of scientific precision and rigor. She followed the company Web site for one year and, during that time, noted problems with individual test results posted to the site. First, some individuals who claimed to have Native American ancestry were actually gauged by tests as having East Asian ancestry instead, which undermines the idea that such categories are genetically distinct. Bolnick also noticed drastic changes in particular individuals' "precise ancestral proportions" over the course of the year despite no official changes being made to the test during that time.[26]

Genetic genealogists have raised similar critiques about the technical limitations of AncestrybyDNA™. During 2005, when I began researching DNAPrint, AncestrybyDNA™ was a daily topic of debate on the Genealogy-DNA Listserv. Lister critiques of AncestrybyDNA™ ranged from forgiving and hopeful that DNAPrint would refine its technique as additional population-level genetic data became available, to more cynical responses. In July 2005, one lister advised someone looking for advice on the best company for Native American DNA testing: "I would stay away from the ethnic ancestry type autosomal tests until we see a valid scientific basis for the results, at present I consider them little better than reading palms or entrails."

Complex or Contradictory Thinking on Race?

The preceding critiques of DNAPrint can, in theory, be addressed by more rigorous scientific practices, such as fuller disclosure of AIMs (although that interferes with commercial interests), and with larger data sets gathered more evenly across the genome and across genetic populations. However, these problems, too, have social and cultural aspects to them. It is both logistically and ethically difficult to sample evenly across populations. Some populations require more work and longer-term relationship building in order to sample. Some populations don't want to be sampled. In addition, samples have been gathered with long-standing racial categories in mind, and not at random or evenly across the groups that researchers do manage to access.

Although many scientists and social scientists today disclaim the possibility of race purity or biological discreteness between populations (although that doesn't mean they have totally given up the ideas), DNAPrint openly promoted popular racial categories as at least partly genetically determined and implied the possibility of purity. The DNAPrint Web site, when it was in existence, referred to individuals of "relatively pure BioGeographical Ancestry" as opposed to "recently admixed peoples." As of June 2005, DNAPrint used interchangeably the phrases "racial mix" and "ancestral proportions."[27] It then explicitly claimed that the test "measures 'the biological or genetic component of race.'"[28] As of July 2009, the Web site, http://www.dnaprint.com/, offered on its "Frequently Asked Questions" (FAQ) page the following definition of BGA: "BioGeographical Ancestry (BGA) is the term given to the biological or genetic component of race. BGA is a simple and objective description of the Ancestral origins of a person, in terms of the major population groups (e.g., Native American, East Asian, European, sub-Saharan African, etc.). BGA estimates are able to represent the mixed nature of many people and populations today."

The company could not help but reinforce the possibility of racial genetic purity when it repeatedly referred to categories commonly understood to be races—European (EU); sub-Saharan African (AF), Native American (NA), and East Asian (EA)—as "major [genetic] population groups." DNAPrint repeatedly referred, for example, on its "Products and Services" page, to the notion of mixedness when it discussed "mixed heritage" or the "mixed nature of many people and populations today." Mixture is predicated on purity.

On its AncestrybyDNA™ home page, under "Products and Services," DNAPrint provided us, as late as July 2009, a visual representation of the notion of population purity. Four prototypical portraits were arranged across the top of the page. They were unlabeled, and indeed there was no need for labels. The first was a person of European ancestry: a fair and slender young woman smiled into the camera. She wore a playful hat to cover her hair. To her right was a person of Native American ancestry: a photo of a youngish man with brown skin and long black hair falling down his back. He smiled, not the typical stoic Indian. Third was a person of African ancestry: a portrait of a dark-skinned (by U.S. standards), handsome, middle-aged black man, who smiled broadly for the camera. Fourth was a person of East Asian ancestry: a middle-aged woman about the same color as the man with Native American ancestry, with well-cut chin-length black hair. She also smiled. In the first moment I laid eyes on this page, I took her to represent Native American ancestry. Then I saw the prototypical Native American–looking male. My initial confusion was fitting. AncestrybyDNA™ sometimes has a hard time telling Native American from East Asian ancestry.

In a move that could dull critiques that the notion of BGA reinforces deterministic biological notions of race, DNAPrint emphasized again on its FAQ page back in August 2009 that in calculating BGA, it was "measuring a person's genetic ancestry and not their race." Yet the "lineages" or "ancestral origins" it purported to measure fall into racial categories that were long-standing before we knew anything about human genetics. On the FAQ page titled "What Is Race?" DNAPrint described recent disciplinary discussions concluding that race is "merely a social construct" as oversimplified: "Over the past few decades there has been a movement in several fields of science to oversimplify the issue declaring that race is 'merely a social construct.' While, indeed this may often be true, depending on what aspect of variation between people one is considering, it is also true that there are biological differences between the populations of the world. One clear example of a biological difference is skin color."[29]

I and others who study race from within the fields of science and technology studies (STS) and bioethics would agree that race as "a social construct" has been oversimplified, but for different reasons.[30] We analyze the concept of race in a genomic era as an object that represents or embodies the co-constitution of natural and social orders. One easy-to-understand example of that is the ways that economic and social disparities have contributed to race-associated medical problems in different racial groups. Take, for example, levels of hypertension, diabetes, heart disease, or prostate cancer that are found to be higher or lower in different racial groups or populations. There are scientists who focus on finding genes that occur in elevated frequencies in some groups that can help explain disease risk or differential responses to medical treatment.[31] Eventually, some researchers hope, such knowledge will lead to cures or treatments.[32] But other researchers emphasize that elevated disease can be tracked to particular histories of deprivation and stress. Such researchers do not view genetic factors as the sole problem, especially if such knowledge does not produce treatment. Rather, they view things like poverty and discrimination as critical factors that lead to health disparities and that society should not neglect in favor of funding sexier genetic research.[33] Some community members share similar worries. I have heard participants at Native American health-research conferences charge funding agencies and scientific communities with focusing too much on genetics. These entities are seen to evade or even compound social problems and historical injustice by making the oppressed or deprived body or population genetically deviant. These issues in research demonstrate the complicated relationship between history, environment, and genomics in understanding the boundaries of race as we use it today.

This type of complexity in understanding race and genomics was not in DNAPrint's purview. Despite its call for "complexity" in thinking about race, DNAPrint relied on overly simplified notions of biology as race and overly simplified notions of science versus society to ground its high-tech research. DNAPrint argued that race is found biologically in "many particular instances of differences between the populations of the world." In support of that claim, the company cited the biological difference of skin color, which it characterized as dramatically different "across populations." (Note the conflation of race and population here, which I'll come back to). DNAPrint focused on skin-color markers as population markers—as race markers.

Yet, as I have mentioned, leading research in the field indicates that the relationship between genes and skin color is not as deterministic as DNAPrint would have us believe. Nina Jablonski, who focuses on the science of skin pigmentation, argues, "The biological basis of skin pigmentation in humans strongly argues against its use as a diagnostic classificatory trait." Skin color varies gradually across space according to specific environmental conditions, not distinctly according to "population" or "race." Thus, it has historically been subject to strong natural selection. As Jablonski puts it, "Pigmentation is a trait determined by the synchronized interaction of various genes with the environment."[34] Near the equator, darker skin is selected for in order to protect against cancer-causing ultraviolet rays. At higher latitudes, lighter skin color is advantageous; it allows in enough UV rays for the human body to sufficiently produce vitamin D, which protects against certain diseases such as rickets. Therefore, populations living in similar environments around the world have independently evolved similar skin pigmentation; witness the similarly dark skin of some people with ancestry on the African and Australian continents. The Eskimo-Aleut are another example of the reason skin color is problematic for classifying race or BGA. Unlike many other peoples living at far-northern latitudes, they evolved dark skin to protect against the high levels of UVA from "direct solar irradiation and reflection from snow and ice." The vitamin D problem has traditionally been solved by their vitamin D–rich diet, and vitamin D deficiencies are manifested as individuals move away from such a diet.[35] Simply put, skin color does not support but confounds the argument that race as a continent-level category has a "genetic component" and that it is a good idea to calculate, with the input of skin-pigmentation markers, biogeographical ancestry according to historically contingent race categories.

Suggesting that there is a "biological basis for race" and using distributions of skin-color markers among different populations to support that statement confuses the logic of the relationship between biology and race. DNAPrint assumed that genetic differences have produced visible differences among people that we can objectively label as "race" differences. (The company also asserted that if society builds important and sometimes discriminatory values and practices around race categories, that is external to the science of genetic race; that produces social race.) DNAPrint's focus on skin-color markers is belied by the most recent science on skin color. Yet the company singles race out as a crucial biological difference among the many that we could use to organize people. DNAPrint, in so doing, ascribed undue importance to skin color. Skin color has long been used to divide the biological continuum of humanity into continental races long before we knew anything about underlying genetic differences. And skin color shapes scientists' choices today about which other biological characteristics to look for. DNAPrint's 175 ancestry-informative markers do not present us with objective scientific evidence that our race boundaries are genetically valid. Markers were chosen with those racial boundaries in mind, and not at random.[36]

Population versus Race in DNAPrint Genomics

The rearticulation of "race" as "population" in the life sciences following World War II was discussed in chapter 1. DNAPrint Genomics did something very interesting in relation to "population" in its own rhetoric about genetics and race. Despite admonishing visitors to its Web site that "race should not be used as a surrogate for population" because it "may lead to over generalization and unfounded stereotypes,"[37] DNAPrint conflated the two notions in its discussion of the connection between race and skin pigmentation. Remember the quotation cited earlier in this chapter: skin color is "one clear example of a biological difference" among different "populations of the world." Therefore, DNAPrint saw skin color as belying the idea that "race is 'merely a social construct.'" DNAPrint, in calculating BGA according to the categories of European, sub-Saharan African, Native American, and East Asian ancestry, clearly drew "populational" lines on a continental level in precisely the same way that human beings have been drawing race lines since at least the sixteenth century. For DNAPrint, continental ancestry equaled population equaled race.

DNA Testing and "Tribe": Markets, Not Markers

I turn now to the concept of the Native American "tribe" and examine the ways in which DNA-testing companies have targeted that category with varying products. I look particularly at GeneTree (and its partner, Sorenson Molecular Genealogy Foundation), Genelex, Niagen, DNA-Today, and Orchid Cellmark. As I did with DNAPrint, I examine in both text and images the technical claims and underlying assumptions about racial or population discreteness that inform company understandings of the concept of tribe. How companies (mis)understand the politics of tribal enrollment and tribal-specific rights to citizenship and resources and how those misconceptions inform the claims they make for their tests are central questions.

GeneTree and the Sorenson Molecular Genealogy Foundation

Since I first started writing and publishing on GeneTree and other DNA-testing companies, in 2003, GeneTree has radically reformatted its Web site and, thus, its marketing output related to Native American DNA. From 2003 to 2006, GeneTree would appear most frequently among DNA-testing companies in Web searches on various aspects of "Native American DNA," and it invariably appeared at the top of the list of search results. The company made itself highly visible to the online community by registering domain names composed of strings of related search terms and then providing links from the corresponding sites directly to GeneTree's main Web site.[38] For example, the URL http://www.americanindiandna.com/ would link directly to GeneTree, thus ensuring that the company appeared frequently on Web searches related to Native American genetic testing. In addition, links to GeneTree appeared most often on Native American–themed Web sites, including Native American genealogy-research Web sites, nonacademic sites concerned with Native American history, and the Web sites of nonrecognized tribes (which are often set up as nonprofit organizations). These links are all inoperative today.

GeneTree has described itself as a "business unit" of Sorenson Genomics, a company founded by Utah billionaire James LeVoy Sorenson. Sorenson, who has been described as "a father of genetic genealogy,"[39] also established a related nonprofit organization, the Sorenson Molecular Genealogy Foundation (SMGF), which has funded a Mormon genetic-research project and private, voluntary data bank at Brigham Young University. (Mormon religious doctrine encourages seeking out ancestors in order to baptize them by proxy. Genealogy research is an important project for the church.)[40] Now described as a collaborator with the SMGF, GeneTree was relaunched in 2007 as a genetic-testing service that offers additional networking and research tools for genetic genealogists.[41] GeneTree is also central to SMGF's global human genetic diversity sampling effort. Volunteers from around the world can purchase a deeply discounted GeneTree genetic-ancestry test (for $49.50 at this writing) in order to get their DNA analyzed and access SMGF's online databases and GeneTree's genetic-genealogy research tools, all while contributing their samples to help SMGF with its global research effort. SMGF notes that it "is committed to building a database that is truly global," and its ultimate goal is to collect one hundred thousand samples.[42]

Despite its global pretentions, SMGF's sampling is geographically highly uneven. The foundation has sampled disproportionately, with the majority of its samples coming from the United States. Of SMGF's nearly ninety-five thousand samples worldwide, 54 percent of those are from the United States;[43] yet the United States has approximately 4.52 percent of the world's population.[44] Additionally, Utah samples account for nearly 25 percent of U.S. samples, making Utah samples account for 12.85 percent of samples globally! The GeneTree-SMGF collaboration, like the Genographic Project, is a fascinating institutional arrangement worth examining elsewhere as an example of the entanglements that can develop between scientific researchers, industry, and consumers, entanglements that pose both risks and potential benefits for ethical and rigorous scientific practice. In this chapter, I focus on GeneTree in the Native American–identity market.

Since its relaunch, in 2007, GeneTree has added to the usual suite of DNA tests (mtDNA, Y-chromosome, autosomal DNA, and DNA-fingerprint tests) a more interactive Web site with tutorials and a glossary on the basic science of genetic ancestry and "molecular genealogy."

GeneTree's revamped "family history networking Web site"[45] offers consumers a Facebook™-type experience dedicated to genetic genealogy. Extended families and "DNA cousins" can network together via the GeneTree site to share not only genetic-sequence information (making use of SMGF's genetic database) but also ancestry documents and family photographs and video; users can also collectively build family-tree charts with genetic relatives far away.[46] The blog and forum features of the site, however, look nearly inactive. That is perhaps not surprising, given other virtual-networking opportunities—both company- and noncompany-specific—that have been available for a decade and have become established.

Since its relaunch, GeneTree has eliminated its blatant focus on Native American–ancestry testing, although it continues to offer testing products that can detect Native American ancestry in mtDNA and Y-chromosome lineages. (It also sold the now-unavailable DNAPrint test.) Interestingly, I spoke on a panel, "Genetic Ancestry Testing: The Public Face of Molecular Anthropology," with SMGF cofounder Scott Woodward at the March 2006 annual meeting of the American Association of Physical Anthropologists. At that time, Woodward was a faculty member at Brigham Young University (BYU). No longer affiliated with BYU, he is executive director and chief scientific officer of SMGF today.[47] During my talk, I levied the same critical analysis of GeneTree and other companies that I apply in this chapter. GeneTree's focus on Native American–ancestry testing was still in full view on the company Web site at that time. By July 1, 2006, the Native American iconography I critique disappeared from GeneTree's Web pages.

An interesting line of research, but one pursued by others, is to investigate, in addition to industry–academia–consumer arrangements, links between this company's demonstrated interest in Native American genetics, its location in Utah, and its close affiliation with Sorenson, a Mormon philanthropist focused on genetic genealogy.[48] The Church of Jesus Christ of Latter-Day Saints has historically taken a special interest not only in the conversion of Native Americans to Mormonism but also more recently in Native American genetic history.[49] Church doctrine asserts that Native Americans are the descendants of Lamanites, a supposed lost tribe of Israel, and are sinners marked by God with dark skin.[50] (Those who were good remained white, "fair and beautiful.")[51] Genetic research into Native American origins has been an important part of genetics work done within and funded by Mormon-affiliated institutions. That GeneTree's commercial decisions have been in some way shaped or prompted by that set of beliefs is a reasonable hypothesis. But I will leave it to other researchers more directly concerned with the relationships between Western religion and science to investigate the connections between GeneTree and Mormon religious preoccupations. I focus instead on the fascinating marketing texts and images that were presented by GeneTree and its spokespeople until very recently across the World Wide Web. GeneTree provides a revealing example of how the notion of "tribe" is understood within broader understandings of the category of Native American "race."

Up until July 2006, GeneTree featured a photograph of a prototypical Native American man with braids under the sign of the DNA double helix, and described the science behind its technoscientific products geared toward answering questions about Native American and other ethnic ancestries: "DNA research on full-blooded indigenous populations, such as *Native American Indian populations*, from all around the world has led to the discovery of *genetic markers that are unique to populations, ethnicity and/or deep ancestral migration patterns*. The markers that have very specific modes of inheritance, and which are relatively unique to specific populations, are used to assess probabilities [*sic*] ancestral relatedness."[52]

Markers are not "unique to specific populations." Rather, ancestry markers of interest are found at higher frequencies in some populations and at lower frequencies in others. The company's claim that markers are unique to populations oversimplifies the relationship between genetics and human sociality. This leads to a second gross oversimplification: GeneTree claimed that it could provide scientifically definitive answers to questions of Native American identity, including tribal-specific identity. In 2003, GeneTree's then–lab manager Lars Mouritsen (chief scientific officer for Sorenson Genomics at this writing) specifically linked DNA markers to tribal groupings: "'Because the database is growing, it's getting more and more sophisticated,' said Mouritsen. 'The more tribes we get, we can really pinpoint which tribe you belong to. Right now it is more localized, but soon we hope to be able to say, based on your maternal heritage, you are part of the Choctaw, or Chippewa, or whatever . . .'"[53]

Mouritsen commits genetic and social-historical oversimplifications simultaneously, because the two cannot be disentangled. It doesn't matter how many samples scientists obtain or how many markers they analyze. A marker for "the tribe" can never be isolated. Why? Any attempt at a simple explanation quickly becomes complicated. First, genetic markers are shared between tribes, because Native American people are not isolated from one another—not prior to colonization and not after it commenced. For example, U.S. Indian policy in the nineteenth and twentieth centuries brought different tribal peoples together in boarding schools and urban Indian-relocation centers. During and since colonization, pan-Indian social and cultural networks have evolved, in part due to federal policy, further aiding Native American genetic and social admixture. Second, because many Native American individuals have ancestry in multiple tribes, they can and do occasionally switch their affiliations from one tribe to another. (Every U.S. tribe I know of has a rule against dual enrollment.) Third, each of the hundreds of U.S. tribes has its own citizenship criteria that change continuously over time with the political and economic winds. Today and in the past, tribes have had standards for who is brought into the fold and who is shut out. (The fact that indigenous self-organization has changed with colonization does not undermine my argument. We may resist the changes wrought by colonization, but the very fact of change is not unique to colonial impositions, and change alone does not invalidate the legitimacy of a tribal entity.) Fourth, individual human bodies that make up the tribe relate to some human bodies and not others. Members of tribes have children, sometimes with other tribal members and sometimes not. Depending on rules of inclusion, such offspring may or may not be considered members of the tribe. They may be considered members of another tribe, or they may not be eligible for enrollment in any tribe. Yet they may nonetheless *identify* as members of that tribe or as Native American more broadly. And finally, "tribes," which consider themselves to be of the same broad historical people, such as the Choctaw or the Chippewa, are actually organized into discrete political entities, giving us multiple "Choctaw" and "Chippewa" tribes, each with its own specific history of enrollment criteria and social and genetic admixtures. What is "the tribe" for the purposes of genetic sampling and testing? The tribe is not, strictly speaking, a genetic population. It is at once a social, legal, and biological formation, with those respective parameters shifting in relation to one another. In order to overlook the genetic complexity of the tribe, the GeneTree spokesperson must simultaneously disregard (or perhaps he does not understand) the highly complex social and political histories and legal mechanisms that have formed Native American racial and tribal formations. Biological and social processes are irrevocably entangled. Simplifying one simplifies the other.

In 2005, Terry Carmichael, GeneTree founder and former company executive, downplayed the links between DNA and tribe when he referred to a controversial aspect of recent tribal-recognition history in the United States. He claimed that people are not interested in Native American–DNA testing "for purposes of claiming rights to a casino." Rather, they simply "want to be able to understand their ancestry a little bit more." He also noted that Native American–DNA testing is growing in popularity more quickly than expected.[54] However, around the same time, Carmichael put forth another point of view in another venue. Carmichael requested that Caribbean Amerindian Centrelink (CAC) Web-site administrators post information on GeneTree's Native American–DNA–testing services. CAC editor Dr. Maximilian Forte, of Concordia University's Department of Sociology and Anthropology in Montreal, wrote the following about GeneTree in his blog: "Terry Carmichael, Vice President for Marketing and Sales at Sorenson Genomics explained in an e-mail that the company 'is providing DNA genotyping services to African Americans and Native American Indians which allows them to trace their roots back to a region in Africa or assess a Native American tribal affiliation.' Sorenson's business units include GeneTree, . . . at http://www.genetree.com/, as well as Relative Genetics at http://www.relativegenetics.com."[55] Careful not to endorse or recommend the services of DNA-testing companies, Professor Forte warned CAC site users of the complicated nature of DNA and Taino identity—that "cultural meanings and practices do not neatly map onto genetic patterns." On the one hand, to be fair, in terms of Taino identity, it may be less troublesome for GeneTree to claim that certain DNA markers help confirm tribal ancestry. The Taino are the only named "tribal" group in existence for those interested in verifying native descent in Puerto Rico. In that case, "tribe" can be used interchangeably with "Native (American)" ancestry in a way that it cannot be used in the mainland

United States. On the other hand, although ancestry-DNA research in Puerto Rico seems to establish that there are lots of descendants of Puerto Rican natives, thus refuting (colonial) historical accounts that generally agree that the native people of Puerto Rico and other areas of the Caribbean were totally wiped out, such DNA analysis does not determine whether an individual is or should be considered a member of a "tribe" called Taino. Forte noted in a personal conversation with me on May 1, 2005, that no Taino organizations at that time required DNA testing for affiliation (they are not federally recognized groups). And he knew of only a couple of individuals who had done such testing out of personal curiosity. The tests are too cost-prohibitive.

By July 1, 2006, GeneTree replaced its contemporary Native American male imagery with a simple black-and-white photograph—nineteenth-century-like—of an aged Native American man in a Central Plains–type headdress. The image was accompanied by the following text:

> Now, with GeneTree's Native American DNA Testing kits, you can discover your roots as they relate to your Native American heritage. Our tests can determine the specific percentage of Native American genetics of the populations that migrated from Asia to inhabit North, South and Central America. It can determine the mix of origin. A specific test can discover whether someone is of 83% Native American descent and 17% European descent. With the new economic opportunities afforded Native Americans, this test can legitimize claims of ancestry, and can even indicate the specific tribe your Native American ancestors belonged to.[56]

In this text, GeneTree dropped its reference to "unique markers." It also softened its language, replacing claims that DNA can "prove" a test taker's Native Americanness or specific tribe with the claim that DNA can "indicate" which tribe one's "ancestors belonged to." Still, GeneTree continued to promote the idea that ancestry markers can link an individual to a specific tribe. It also added a new idea to its repertoire of unwarranted claims. It linked the presence of DNA markers to the right to invoke economic benefits reserved in treaties and law for Native American tribes and their citizens. It is not actually the case at present that DNA markers are evidence for making such claims. The question is, will this brand of gene fetishism actually come to inform the way laws, policies, and provisions for Native American economic development evolve? Remember the DNAPrint customer who used a trial version of AncestrybyDNA™ to "prove he qualified for a business venture exclusive to Native Americans." That outcome was not only reported in the press but also noted on DNAPrint's own Web site.[57]

Genelex and Niagen: Marketing Tribal Rights via Markers

A third company, Genelex, warrants attention because it has marketed its services in forums that are the domain of federally recognized tribes. Although the company now focuses on familial relationships via DNA-profile tests, several years ago it was regularly advertising in the prominent weekly *Indian Country Today* and in the *Navajo Times*. The company's ads were both technically problematic and at odds with history and with contemporary governance practice. For example, in the September 22, 2004, *Indian Country Today*, a Genelex ad proposed to readers the following: "Do you need to confirm that you are of Native American descent? Recent advances in genetic testing have put the answer to this question at your fingertips. Whether your goal is to assist in validating your eligibility for government entitlements such as Native American Rights or just to satisfy your curiosity, our Ancestry DNA test is the only scientifically rigorous method available for this purpose in existence today." Like GeneTree, Genelex also claimed that its "Ancestry DNA test" revealed genetic markers that are "unique to Native Americans."[58]

In the same light, another company, Niagen (now out of business), made similarly strong and controversial claims about the applicability of Native American–DNA testing to Native American and tribal-specific identity. Its Web page for "native Indian heritage testing" claimed the following:

> Native Indian heritage testing are [*sic*] for those individuals who would like to determine the presence of genetic native Indian ancestry. Utilizing the latest in DNA research individuals are able to determine the percentage of native Indian ancestry within their genetic line. Such testing is essential for documented proof of nativeness and use in government or other benefits.
>
> Native Indian heritage DNA testing determines the ancestry from most tribal origins and is a highly accurate means of determine

[*sic*] percentage or full native Indian ancestry. Utilizing researched gene specific loci, specific allelic markers common within Native Indian genotypes are measured to determine the degree of Native Indian Ancestry.[59]

Several of Genelex's and Niagen's claims demonstrate how some purveyors of genetic tests scientifically mischaracterize Native American DNA and suggest its application in controversial ways. Genelex claimed that ancestry-DNA tests, whether mtDNA or Y-chromosome (it sells both), identify uniquely Native American markers.[60] The claim is not supported by the science. "Native American" mtDNA markers, the mutations that characterize haplogroups A, B, C, D, and X, are not, in fact, found uniquely in Native Americans.[61] They are found most often in Native Americans, but because of migration histories they are also found in other populations in smaller percentages.

Y-chromosome markers such as M19 and M3 are not uniquely Native American, either. Not all Native American–identified males have them, and there are certainly non–Native American–identified males who do have them, which is not surprising, given how populations admix. At any rate, there has not been widespread sampling for Native American markers of any type. Regarding the AncestrybyDNA™ test (which Genelex and Niagen also marketed, respectively, under the headings "Native American Ancestry DNA Testing" and "Native Indian Heritage Testing"), so-called Native American markers surveyed in that test are also not unique to Native Americans. Although some of the genetic markers used in mtDNA, Y-chromosome, and DNAPrint's autosomal-marker test have, to date, been found only in individuals already identified as Native American, neither Natives nor non–Native Americans—no matter how one delineates such samples—have been systematically sampled. We do not have an in-depth understanding, based on defensible sample size, of how such markers are distributed globally.

Thus, it makes little sense for scientists and companies to speak of such markers as found "uniquely" in Native Americans. Given what I have said about the impossibility of determining tribe—in either the past or the present—via genetic markers, it was also not accurate for Niagen to speak of "Native Indian heritage DNA testing" as determining "ancestry from most tribal origins." Even though some non–Native American test takers certainly find so-called Native American markers in their genomes, the language of these companies implied that such markers are definitive of racial and/or tribal identity. Despite claims about "scientifically rigorous methods," these are not scientifically rigorous ideas. In addition, they have important political implications.

Companies such as Genelex and Niagen have exaggerated what DNA tells us about ancestry and its correspondence with Native American racial, ethnic, and tribal identity. That said, I do not charge such companies with willfully misleading the public. Their claims demonstrate that genetic science does not exist in a cultural vacuum. Genelex's and Niagen's language reveals an idea that persists unevenly within and without genetic science: that genetics relates to ethnic/racial group identity in a deterministic way. Thus, tests for genetic markers are sold as proof or validation of particular identities. In such cases, companies such as these promote understandings of ancestry DNA in which it obfuscates and stands in for legal and social practices and political histories that have constituted Native American identity in its tribal and racial forms. Both Genelex and Niagen have engaged in gene fetishism.

DNAToday and Orchid Cellmark

DNAToday and Orchid Cellmark are two companies that have specifically marketed the DNA-profile, or parentage, test specifically to U.S. tribes and Canadian First Nations for purposes of enrollment. The parentage test is both more and less informative than the genetic-ancestry tests discussed so far, because it gauges relatedness at a different biological level. DNA profiling can be used to confirm close relations (such as mother, father, and siblings) with very high degrees of probability. It does not analyze markers judged to inform one's "ethnic" ancestry. By contrast, Native American–ancestry DNA tracing shows that one shares relatively few, often noncoding markers and thus probable but more distant relatedness with historical Native American populations.

The DNA profile is commonly used as a paternity test, although it can also be used to examine maternal lineages. The DNA profile is also commonly used in criminal cases—to prove, for example, that a strand of hair or skin cells found on a crime victim belong to an individual suspect. The technology examines repeated sequences of nucleotides such as GAGAGA, called "short tandem repeats" (STRs). We inherit STRs

from both parents, but our total individual STR pattern is, in practical terms, unique. DNA fingerprinting examines very specific patterns of multiple STRs. A single sequence is not unique. But when viewed in combination with other STRs, the pattern becomes increasingly distinctive; only 1 in 60 million individuals might exhibit such a pattern.

Many companies sell this technology to the public and to other governmental agencies, and most do not expressly target tribal enrollment. Inaccurate knowledge of biological parentage is not a principal problem in enrollment today. But it is exactly because companies market technical information that is often redundant that they make an interesting case for analysis. The cultural capital of "genes" and "DNA" and the zeal in the marketplace for personalized DNA services suggest that these companies market the idea of rigorous, scientific knowledge as much as they do the technical information to be had from a test.

One ancestry-DNA-testing company scientist has expressed dismay at the marketing of parentage tests at genealogy conferences to those interested in Native American ancestry.[62] He felt that selling paternity tests in this market was intended to mislead consumers by not distinguishing between parentage tests and the "superior" technical aspects for Native American genealogical research of Native American marker tests. However, U.S. tribes and Canadian First Nations are a unique subset of the DNA-testing market. Genetic genealogists who are interested in making genetic links between individuals with the same surnames and sometimes deep genealogical histories find ancestry-DNA tests obviously more useful for that activity. But U.S. tribes and Canadian First Nations are concerned with who gets enrolled. The parentage test precisely demonstrates recent biological ancestry and relationships that must be documented for conferring tribal membership.

Given current tribal-enrollment rules, genetic-ancestry tests and the forms of biological relatedness documented by genetic-ancestry mtDNA, Y-chromosome, and autosomal analyses are not the same forms of biological relatedness with which tribal-enrollment offices are concerned. Genetic-ancestry test findings do not generally illuminate an applicant's close biological relationship to an individual who is or was already on the tribal rolls or to an individual who was on the tribe's "base rolls." Genetic-ancestry tests document the fact that a test taker is descended from unnamed "founding ancestors" who first settled the Americas.

Such genetic information is interesting, but at this point it is irrelevant from a tribal-enrollment point of view.

By contrast, parentage tests might be useful in certain individual enrollment cases. They are already commonly used when an applicant's biological parentage is in question and a parent's tribal status is essential for the applicant to be enrolled. However, in the majority of tribal-enrollment cases, biological parentage is not in question. Theoretically, then, parentage testing would be used only occasionally. But because of an onslaught of enrollment applications (this is especially true of wealthier tribes), we see tribes increasingly move to a system in which a DNA-parentage test is required for all new applicants, and is even sometimes required of existing tribal members under new rules for across-the-membership testing. This is done in part to reduce the work burden of tribal-enrollment offices related to managing large volumes of identity documents. Of course, there are risks to this approach, including finding "false" biological parentage where it did not exist in the first place. As an alternative approach, my own tribe will accept affidavits from three enrolled relatives testifying as to their relationships to an applicant for the purpose of using parental lineage for enrollment. In this case, familial testimony and relationships are not trumped by genetic tests.[63]

OriginsToday™ and the Tribal Identity Enrollment System™

DNAToday, until it declared bankruptcy, in 2006, marketed the parentage test in combination with its Tribal Identity Enrollment System (TIES), which included an ID card and a "smart card" with an individual's photograph and embedded computer chip that could hold a "DNA profile and an individual's other tribal government records" (such as enrollment, voter-registration, and health- and social-services data). Such records would be managed with the Origins Today™ Sovereignty and Sovereignty Plus Edition software.[64] The same software is now marketed by Dynamic ID Solutions,[65] with Steven Whitehead, former DNAToday president, at the helm. Tribal-government customers of Origins Today span the country and include tribes in Oregon, Washington State, Arizona, Wyoming, Nebraska, Oklahoma, Minnesota, Mississippi, and New York State.[66] This technologically savvy suite of products aims to "factor out political issues" and establish "clear answers" for "all future generations."[67] Thus, tribal identity can be technologized and made more

precise by the merging of two types of information: computer code and genetic data. The software package Origins Today™ sports a DNA double helix within its logo.[68] Thus, the "information" and "code" concepts are deployed both literally and metaphorically.

I first encountered DNAToday at a 2003 national tribal-enrollment conference attended by approximately two hundred individuals at a cost of $459 each. I was informed about the conference by the enrollment director of the tribe in which I am enrolled. Coincidentally, DNAToday had already approached our tribe to pitch its services and was declined. The conference host, DCI America, allowed me to enroll—at full fee—as a graduate student who was studying the role of DNA testing in Native American identity. DCI America, which advertises itself as "the premier Native American and First Nation training organization in Indian Country," hosted the conference in the ballroom of an upscale corporate chain hotel in New Orleans.[69] When I entered the brightly lit ballroom on that October morning, with its tapestries and sconces on the walls and its massive chandeliers, the large circular tables were already full with conference-goers, eight to a table. I later learned that tables were largely populated by multiple staff members of single enrollment departments. Attending were tribal-enrollment officers from tribes all over the United States, from the West Coast, the Southwest, throughout the Midwest, and the East. Arriving too late to grab a seat at a table (and aware that my presence might inhibit the potentially confidential conversations of tribal-enrollment staff), I spotted an empty chair in a row of chairs lined up against the wall of the ballroom next to an electrical outlet. I sat down next to several individuals who I later learned were DCI America staff and interns. Dressed in business attire, almost none of them appeared to be Native American, unlike the sea of brown faces—many middle-aged and elderly women—at the conference tables. As I pulled out my laptop and plugged it in for note taking, I drew curious looks from the staff.

DCI America is one of several for-profit training and technical-assistance companies operating in Indian Country. Geared toward the directors and staff of federally recognized tribal-enrollment departments, its conference is held annually at different locations around the United States. Billed as addressing the inefficiencies of tribal-enrollment procedures, in 2003 it played out like a three-day infomercial for DNA-testing services. DNAToday was featured front and center, leaving the impression that genetic testing is key to solving existing inefficiencies in enrollment. DNAToday then–company president Steven Whitehead (a self-declared former insurance salesman) sat on the keynote panel in the conference's opening session and took all questions related to the use of DNA in enrollment, giving the impression that the genetic science at play somehow involves only questions of genetics and not the politics and histories of enrollment. He was accompanied by a tribal chairman, a tribal court judge, and a Bureau of Indian Affairs field-office superintendent who addressed questions of law and policy with no reference to genetics.

In addition, DNAToday scientists and software technicians staffed a trade-show-like table in the conference registration area, where they offered a product demonstration. Company executives and scientists were given the title of instructor. DNAToday hosted "workshops" throughout the three-day event, some of which played out more like glitzy product demonstrations. At one point, a conference attendee was chosen, through a raffle, to receive a free DNA test. As her name was called out, the room erupted in applause. The conference-goer walked to the front of the ballroom and sat down at a table. Whitehead, speaking into a microphone so that he could be heard in the cavernous venue, explained that she and a witness were signing an informed-consent form. Whitehead then narrated the taking of her DNA in a hushed voice into his microphone, almost as if he were a sportscaster narrating a quiet, tense moment in a sports match: "He is swabbing one side [of her cheek], now he's swabbing the other."

Other sessions led by DNAToday amounted to extended sales pitches for the technologies and services offered for sale, not unlike presentations one would see at a trade show. Workshop overhead projections included "Reasons to Select the T.I.E.S.™ System," "Your Investment Options," "3 Ways to Get Started," and the name of a contact for further information.[70] DCI America not only hosted DNAToday as a presenter in this annual conference, it also actively marketed the DNAToday suite of products. The president of DCI America presented a quick overview of the OriginsToday™ software during the tribal-enrollment workshop, and the organization endorsed the software on its Web site. During a break, one tribal-enrollment director turned to me. (I had found a seat at a table because of dwindling workshop attendance on that lovely fall afternoon in New Orleans.) She commented that one of the other popular training organizations in Indian Country also led a tribal-enrollment workshop, but its facilitators were "more like BIA people." By this she meant that they focused more on the intricacies of enrollment according to federal and tribal government procedures, whereas at this conference DNA testing was discussed as a potentially significant component of enrollment.[71]

Despite the focus on DNA-testing services, conference organizers did not offer basic genetics education, especially as it might relate to current tribal blood rules. It was clearly necessary. One attendee asked whether DNA testing could be used to scientifically determine the blood quantum of tribal members. DNAToday president Whitehead responded that DNA testing says nothing about blood quantum, "but you can prove a child is related to a mother."[72] Perhaps clarifying relationships of DNA to symbolic blood was never an aim. Were those complex and messy blood rules and meanings viewed as the very problems to be solved with a more straightforward and technologically demonstrable biological relationship? It is true that Whitehead clarified that the "paternity test" proves a biological relationship that consequently allows a tribe to confidently use parents' blood-quantum calculations to determine those of offspring. But during the keynote panel, he also claimed that DNAToday's technology is "100 percent reliable in terms of creating accurate answers" to questions of tribal enrollment.[73] Whitehead's responses were almost non sequiturs. He did not comment on the loaded and very complicated symbolic meanings ascribed to blood that go beyond the lineal biological relationships that DNAToday and other companies focus on. He took the blood-quantum question as purely a molecular-knowledge question.

Indeed, the marketing of parentage tests as a definitive solution to the troublesome politics of enrollment demonstrates a blunt edge to company analyses of contemporary tribal-enrollment debates. In another session, Whitehead advocated tribal-wide paternity testing so that "only Native Americans who deserve to be members of your tribe will be."[74] Yet another company's spokesperson has claimed that a DNA-parentage test "validates tribal identity" or "can preserve a tribe's heritage."[75] These statements gloss over much more complex processes in which tribal communities entertain deep philosophical and political-economic disagreements about who should rightly be a citizen and who should not, who belongs and who does not. First, tribes must privilege certain relationships in their definition of what constitutes the tribe. Kirsty Gover has shown that those relations in the twentieth century became increasingly biological or genealogical and were less based on residence, marriage, or adoption.[76] In some cases, next, a DNA test can be used to confirm or disavow the emphasized biological relationships. But just as the definitions of the emphasized relationships have changed over time, they will continue to change along with demographic change and policy upheavals. The boundaries and definitions of the "tribe" will continue to change as a result. By glossing over the initial decision that tribes make about which relationship(s) to count as pivotal, DNA spokespeople fetishize molecules, making certain shared nucleotide sequences stand for a much more complex decision-making process. They make the DNA test appear to be a scientifically precise and universal answer instead of a technique chosen as part of a broader set of political maneuvers over which there are deep philosophical, cultural, and economic disagreements within tribes—disagreements about who should count as, for example, Dakota or Pequot or Cree.

Several times during sessions at the 2003 conference, it became clear that some (perhaps many) conference-goers did not understand the difference between the DNA profile that DNAToday sells and genetic-ancestry tests that other companies sell. To be fair, Steven Whitehead in those moments clarified effectively the problem with some genetic-ancestry-testing company statements, namely, their inaccurate implication that certain ancestry markers can prove particular tribal origins: "DNA is very precise in determining relatedness, but it can't make distinctions regarding ethnicity." He explained further that "there is no gene specificity to a particular tribe or to blood [quantum]." At another point, he stated, "There is no known technology that can identify you as an Indian, no known technology that can identify you as a tribal member." But Whitehead's clarity on this point was contradicted by the way he glossed over the very narrow technical product of the technology he sells, the DNA profile, which in practice gets inserted into longstanding, politically contentious, and complex enrollment processes. Instead, Whitehead emphasized what seemed to him to be obvious links between "who's entitled to be enrolled and not enrolled" and particular genetic relations ("It [DNA] can so clearly define relationship ties"). This is a genetically fetishistic position, making DNA stand for a much more complex set of relationships and processes. Although there was lots of talk about tribal sovereignty and decision making, the DNA profile was represented as more than a discrete tool to be inserted by a tribal council into a much larger and complex process; the DNA profile was represented as a key and scientifically rigorous undertaking "that will preserve culture."[77]

In 2003, DNAToday offered its "legal" paternity test for a group of two to three individuals (an individual plus one or both biological parents) for $495.[78] The price of such services remained the same through 2010.[79] Take a ten-thousand-member tribe. The number of tests required to retroactively test every member of that tribe might be four thousand (an average of 2.5 people included per test). At $495 per test, the cost to test all members would be nearly $2 million. Add to that the costly ID card sold to accompany tribal-wide DNA testing. DNAToday generated each individual card in its facilities at a cost of $320 per tribal citizen. Thus, it would cost $3.2 million for a ten-thousand-member tribe to equip each tribal member with a programmed card. In addition, the tribe would need to return to DNAToday to purchase replacement cards, and the purchasing tribe would have the near-impossible logistical task of testing all tribal members, including those living off the reservation.

Such products and services also present concerns for tribes related to privacy and legal self-determination. For example, although the tribe purchasing the DNAToday products endorsed by DCI America would determine the specific data to be programmed to cards and would maintain the database, DNAToday would retain control of issuing original and replacement cards. DNAToday did note that it purged data from its system after generating cards, and a tribe's confidentiality agreement would no doubt require such safeguards. But issues of property and control run deep in this domain. DNAToday would store tribal DNA samples, which also raises privacy concerns. Whitehead responded during a question-and-answer period to a question about privacy, noting that after identity markers were analyzed, individuals could request that their samples be destroyed and receive affidavits certifying as much. Enrollment applicants could also sign as "registered agents" on their samples, which would then require that they issue notarized permission for others to obtain access to their DNA. But as scholars of biological property have demonstrated, the law has not in fact kept up with technologies and the array of potential privacy risks that emerge.

Given privacy concerns and given that DNA storage is fairly low-tech and, in relative terms, not too expensive (blood samples can require a freezer set at -80 degree Celsius, which costs $5,000–$10,000, whereas a DNA sample that has been chemically preserved can be stored at room temperature), tribes would probably want some form of tribally controlled management of samples. To do otherwise would give non-tribal, for-profit companies too much control over tribal members' valuable biological data. I raised these issues during the question-and-answer period of a DNAToday workshop. The company representative's response was again a non sequitur—simply that the industry standard was to include twenty-five years' storage at no extra cost, so why not just let the company handle it? For the DNAToday executive, technology and cost were the golden keys to solving the most pressing issues of sovereignty in Indian Country.

There were a few individual tribal-enrollment officers and one tribal-community panelist who expressed hesitation about DNA testing and the smart-card system, either because they seemed to contradict notions of tribal citizenship based on indigenous cultural concepts or because the technologies seemed invasive. One elderly woman in the audience was clearly irritated when Steven Whitehead of DNAToday explained how far the smart card and its attendant software could be made to reach across tribal programs and facilities to monitor individual tribal-member engagements with different programs, their records, histories, and identities. She asked Whitehead, "Does anyone else carry around this type of info? Do you?" I took her to mean, "Do we expect non-Natives to be so monitored in their day-to-day lives?" DNAToday noted that it provides the same technologies to the U.S. State Department and the Department of Homeland Security. Question answered. But the vast majority of conference attendees seemed either unworried or unimpressed by the technology or hopeful that it would provide a scientific solution to their considerable and messy political problems in the enrollment office.

In October 2010, I returned to the DCI America National Tribal Enrollment Conference, this time hosted at the sleek, tribally owned Hard Rock Casino and Resort Hotel in Albuquerque, New Mexico, with—to my great pleasure—a coffee bar staffed by an efficient and capable Native American barista. I was in cultural-hybridity heaven. This time, I was an invited speaker on a panel presentation on DNA and tribal enrollment. I was grateful to be given the opportunity to talk to a tribal audience about the differences between genetic-ancestry testing and the DNA profile as they relate to tribal enrollment and, more broadly, to tribal sovereignty. This perspective had been absent from the 2003 tribal-enrollment conference. My presentation on the panel was followed by that of a DNA-testing-company executive, this time from the company DNA Diagnostics Center (DDC), which sells the same type of tests as those marketed by DNAToday and Orchid Cellmark. DDC also had a booth outside the main conference hall advertising and explaining its services.[80] In 2010, however, the role of DDC was much scaled back from the role that DNAToday played at the 2003 workshop. Still, as in 2003, basic genetics education was necessary. Before our DNA-testing panel, a participant—a tribal-enrollment-office staff member—asked of another representative from a tribe that does DNA-parentage testing, "How does DNA work? Do they determine if your DNA is from a particular tribe?" The representative from the DNA-testing tribe responded, "No. It just says 'Indian.'" Ouch.

During this conference, the DDC spokesperson did a quick, efficient overview of the basic science. I also covered some basic science in my talk. I hope it helped. For those not in the know, the enrollment staff person from the DNA-testing tribe communicated a wholly inadequate understanding of what DNA profiles reveal. That communication, combined with imprecise company language—DNA tests "validate tribal identity" or "can preserve a tribe's heritage"—conveyed the idea that DNA testing does indeed reveal Native American–specific genetic material, that Native American race or tribe can be rendered a genetic category. Again, as in the 2003 DNAToday presentations, DDC tried to be clear about the basic science. In fact, the DDC spokesperson was even clearer, devoting much more presentation time to the science than DNAToday had done. He largely pitched his company's DNA-testing services outside of the main conference sessions, at a display table outside the ballroom. Yet the DDC representative also contradicted his careful elucidation of the science with genetically fetishistic statements that seemed to condition tribal identity on a DNA test. And indeed, there was confusion displayed in conversations between different enrollment staff during question-and-answer sessions about exactly what the DNA tests reveal about genetic relationships and, by extension, about Native American or tribal-specific identity.

Unlike in 2003, multiple tribal-enrollment officers at the 2010 conference noted their tribes' use of DNA-parentage testing for enrollment. Yet, interestingly, symbolic blood loomed much larger than gene talk in participants' discussions of tribal-enrollment processes and debates. And unlike those with genetics-based knowledge, the officers who spoke out loud were expert in their discussions about the nuances of blood rules in enrollment. Just days before attending the conference, I had written about Kirsty Gover's article "Genealogy as Continuity" for another chapter of this book. Listening to participants' comments over the two days, I heard confirmed Gover's insights about the turn to genealogical tribe making. Any talk I heard of genetics at this conference showed that DNA profiles are clearly supportive of the move to constitute the genealogical tribe. But tribal-enrollment directors and staff spoke at greater length about (hoped-for) links between higher blood quantum and cultural affiliation. They recounted tribal community debates about whether to retain or lower blood-quantum requirements. Some talked about the turn to lineal-descent criteria instead, but they worried then about the enrollment of too many individuals who really had no cultural connection to their tribes. In addition, several participants expressed much concern about increasing numbers of reservation residents, including children, who cannot be enrolled because of having too many tribal lineages in their backgrounds; they cannot meet the blood-quantum requirements of any tribe from which they are descended. As a result, they are disenfranchised when they actually do live within and are connected culturally and socially to the tribal community.

Precisely the same predicaments had been recounted by tribal-enrollment staff during the 2003 enrollment conference. Nothing had changed on that front. Yet DNA testing was clearly of greater interest in 2010 than it had been in 2003. A poll taken by a show of hands on the last day of the 2003 conference showed that all except ten participants had no opinion on how useful or important DNA would become in enrollment. At my breakout table, the women I sat with preferred to chat rather than engage deeply in the role-playing exercise of constructing a DNA-enrollment ordinance. Yet even in 2010, amid increasing DNA testing by tribes, the use of the DNA profile continued to support a broader understanding of relatedness as figured through symbolic blood. The question now seemed to be less about how "useful" DNA would become in enrollment; it's on the rise. Rather, the question seemed to be, first, For how long will blood continue to dominate in the tribal imaginary and condition the adoption of DNA tests? and, second, How will blood and genetic concepts increasingly work together to (re)constitute the notion of the tribe?

Orchid Cellmark: Which Truth Is in the DNA?

In Orchid Cellmark (or Orchid) advertisements that ran from May through November 2005 in a national Native American newsmagazine targeted to tribal governments, *American Indian Report*, an Indian in silhouette faced the setting sun. The image, in purples, pinks, and blues, recalled the classic "End of the Trail" depiction, the broken nineteenth-century Indian on horseback. But Orchid's Indian sat upright. Did he envision a more hopeful future for his progeny than had his nineteenth-century counterpart? After centuries of predicting their demise, could scientists now testify to the American Indians' survival? Orchid's ad announced that genomic technologies reveal "the truth." Were readers to assume the ad referred simply to DNA markers, or were they meant to think that Orchid's science can reveal—via those markers—an ethnic essence, the stereotypical nobility and tradition portrayed in the image?

In a unique move, Orchid Cellmark entered the Native American–identity market offering a more complete array of DNA-testing services than did other companies. Like DNAToday, Orchid promotes the standard parentage test for U.S. tribal and Canadian First Nations enrollment.[81] But it also markets tests for Y-chromosome and mtDNA Native American markers, as well as—when it was available—the DNAPrint test that supposedly determined one's "percentage of Native American–associated DNA."[82] By offering both the DNA-fingerprint and the genetic-ancestry tests to the Native American–identity market, Orchid Cellmark presents a more complex set of implications in discussions about Native American identity. Like DNAToday, Orchid stresses that its DNA-parentage test can "confirm the familial relationship of specific individuals to existing tribal members."[83] Again, the parentage test is technically helpful for tribal enrollment in a minority of cases, when a biological relationship with named individuals is in doubt and proof of that relationship will support additional enrollment criteria. But many people with demonstrably close biological relationships with tribal members are still ineligible for enrollment, because they do not meet subsequent nongenetic criteria.

Native American–marker tests are bad technical matches for tribal enrollment. They do not confirm enrollment applicants' relationships with named ancestors who were also documented tribal members, the central requirement in all tribes' enrollment processes. A company spokesperson confirmed in a 2005 interview that U.S. tribes and First Nations had purchased only paternity tests for enrollment purposes.[84] However, genetic-ancestry tests are important symbolically. They promote the notion of shared ethnic/racial genetic material by constituting "Native American," "European," "African," and "East Asian" DNA. The AncestrybyDNA™ test also provides a percentage calculation of "biogeographical ancestry" allocated between such categories.

It is problematic that Orchid Cellmark markets ancestry tests while also actively building a customer base of tribal and First Nation governments (not simply Native Americans). It would not be such a stretch to embrace the notions of ethnic/racial DNA markers and biogeographical ancestry percentages (if the latter are reincarnated in a company structure other than DNAPrint), especially for those tribes that gauge relatedness to group by counting the number of one's specifically named and enrolled ancestors in order to calculate a blood-quantum fraction. If tribes begin to think in terms of combining parentage with genetic-ancestry tests, which geneticize the symbolic blood-quantum fraction, Orchid Cellmark has a package that, although questionable in its technical applicability to current enrollment standards, has big symbolic value. It just might appeal to tribes, especially those with contentious enrollment processes (such as wealthy gaming tribes flooded with enrollment applications from would-be beneficiaries). For better and for worse, Orchid Cellmark demonstrates a more complicated understanding of the cultural politics of contemporary Native American identity. Where other companies target either "race" or "tribe" more strongly, Orchid Cellmark's array of products accounts for Native American identity as both racial and tribal.

The names "tribe" (or First Nation) and "race" contradict and overlap each other as they help construct the different Native American identities to which DNA-testing companies respond. Advocates for tribal-government sovereignty typically talk about the fundamental difference between the concept of race and that of the tribe/First Nation in justifying self-governance. They commonly argue that tribes are nations (also implying that they have a distinct culture), and not races. And nations decide who gets citizenship and who does not. Yet tribe and race share common ground. Part of that common ground is symbolic blood, which is increasingly complemented by DNA, although blood and gene talk take different forms when used to support different positions (such as reckoning race versus reckoning tribal membership). Orchid Cellmark appeals to DNA simultaneously as an indicator of race and as a representation of one's relatedness to tribe. Because tribes operate in a world that is both tribalized and racialized, they struggle to mediate a Native American identity according to those sometimes-contradictory, sometimes-overlapping categories.

More than that of any other company discussed, the work of Orchid Cellmark reflects the breadth of ways in which DNA testing and gene talk can expand into territory previously claimed by blood-quantum regulations and symbolic blood. Unlike blood, DNA testing has the advantage of claims to scientific precision and objectivity. Orchid Cellmark's director of North American marketing has noted that in using DNA-fingerprint analysis, there is "no possibility of incorporating a subjective decision into whether someone becomes a member or not."[85] Of course, whether or not someone is verifiable biological kin of the type indicated by a parentage test is not an "objective" enrollment criterion. Allowing a DNA profile to trump other ways of reckoning kin for purposes of enrollment (such as using blood quantum as a proxy for cultural affiliation by counting relatives, or obtaining a signed affidavit of family relatedness) prioritizes technoscientific knowledge of certain relations over other types of knowledge about the same and other relationships. Nonetheless, the idea of scientific definitiveness attached to genetic testing is influential, even if it is not realized. Thus, the DNA profile may increasingly look like a good complement to traditional blood (quantum) and other nongenetic documentation—especially if traditional documentation of named relations is difficult to obtain or if enrollment applications are politically and economically contentious. In addition, the increasing use of the DNA profile in concert with existing blood rules may condition tribes' eventual acceptance of DNA knowledge as a substitute for tracking blood relations. Some will see such a move as advantageous, as scientifically objective and less open to political maneuvering. Yet DNA testing will not solve what is the most crucial and divisive problem in contemporary enrollment debates: in the majority of cases parentage is not in question, but because of outmarriage, increasing numbers of tribal members' offspring cannot meet blood requirements. They simply do not have enough sufficiently "blooded" parents and grandparents to meet the standards set by tribes. Using DNA-parentage tests can be seen, on the one hand, as supporting Gover's genealogical tribe, in which blood rules move away from the broader racial category of "Indian blood" to constitute blood as tribal-specific. In that move, she argues, tribes are relying *less* on race than have federal agents in the past.[86] On the other hand, the increasing tribal practice of DNA testing across the membership can also pave the way for a reracialization of Native Americans by promoting the idea that the tribe is a genetic population. The incommensurable nature of the DNA profile with the genetic concept of population and its re-articulations of older notions of race will be lost on many observers. In addition, if genetic-ancestry tests come to be coupled with the DNA profile—the DDC spokesperson at the 2010 tribal-enrollment conference noted the use sometimes of mtDNA-lineage tests to ascertain maternal lineages in tribal-enrollment cases—"race" is certain to loom larger in our conception of Native American tribal and First Nations identity in the United States and Canada.

If race in the form of genetic ancestry comes to have greater influence in how we understand Native American or Aboriginal identity, historical and legal relationships and federal-government obligations to tribes and First Nations may fade further from view. In turn, claims to land and self-governance may be denied or justified by the absence or presence of Native American DNA in individual claimants. This may have two effects. First, anti-indigenous interests will have strong ammunition to use against tribes which they already view as beneficiaries not of treaty payments but of special, race-based rights. Second, groups without historical-colonial relationships, heretofore racially identified as other than Native American, may increasingly claim indigenous-nation authority and land based on DNA. The marketing of DNA tests to tribes, First Nations, and others clearly has much broader implications than simple revisions in enrollment policy.

Given the commercialization of genetic-ethnic identities and capitalism's ubiquitous reach, we must consider that the scientific object of Native American DNA (or the definitive absence of its markers) will be important in remaking Native American identity in the twenty-first century. DNA markers quite simply will be very valuable in certain quarters for making or refuting claims of Native American identity that otherwise might be very difficult to make or refute. Thus, DNA markers—when there is something tangible to gain—may be used to legitimate claims that contradict and potentially contravene prior tribal claims based in historical treaties, law, and policy, even if the groups that use Native American–DNA analysis do not intend to undermine existing tribal claims and law.

In October 2010, I left the tribal-enrollment conference in Albuquerque thinking that Native Americans selling cappuccinos, serving slightly upscale steak dinners, and renting hotel rooms decorated in a tranquil and vaguely Asian theme seemed a desirable kind of cultural hybridity (I can't comment on the gaming facilities; I prefer café Wi-Fi to slot machines). Of course, gaming is central to the constitution of the twenty-first-century capitalist Indian, and its wealth challenges well-worn ideas of authentic indigeneity.[87] Still, the prospect of wealthy gaming tribes and DNA-testing-company executives reconfiguring the tribe or First Nation into a genetic entity feels particularly troublesome, like a culminating act in the unsettling-from-within of the very grounds for claims to indigenous sovereignty.

3

GENETIC GENEALOGY ONLINE

> In all of us, there is a hunger, bone-marrow deep, to know our heritage—to know who we are and where we have come from. Without this enriching knowledge, there is a hollow yearning. No matter what our entertainments in life, there is still a vacuum, and emptiness, and the most disquieting loneliness.
>
> —Alex Haley, quoted in *Trace Your Roots with DNA*, by Megan Smolenyak Smolenyak and Ann Turner

> Biology is not the body itself but a discourse on the body . . . a linguistic sign for a complex structure of belief and practice through which I and many of my fellow citizens organize a great deal of life.
>
> —Donna Haraway, *Modest_Witness@Second_Millennium.FemaleMan©_Meets_OncoMouse*™

Genealogy research is perhaps the most popular U.S. American pastime.[1] This chapter explores the practice of "genetic genealogy," or genealogical ("family tree") research that makes use of ancestry-DNA tests to fill in documentary gaps. Often called an obsession, genealogy research had an estimated forty million practitioners in the United States alone by 2004.[2] Today, Rootsweb.com boasts more than thirty thousand genealogical mailing lists.[3] During the first decade of the twenty-first century, the genetic-testing phenomenon grew quickly among genealogists.[4] Between 2001 and 2006, we know that nearly five hundred thousand people purchased tests.[5] At an estimated rate of growth of 80,000 to 100,000 tests sold per year, noted genetic genealogist Blaine Bettinger predicted that the number could hit 1 million by 2009.[6] That suggests 1.3 million tests sold by 2011. In 2005, professional genealogist Megan Smolenyak Smolenyak surveyed readers of *Ancestry Daily News* about their use of genetic-ancestry tests and found that 25.4 percent of survey respondents had already engaged in genetic-genealogy testing. Admitting the bias of her survey, in which those who had engaged in such testing would be more likely to respond, she was nonetheless surprised at the numbers. Of her respondents, 14.4 percent had taken two or more tests.[7] DNA-testing companies play to this market with stereotypical race imagery and overly simplified scientific claims. But do genealogists—a significant piece of the market—share those understandings?

Many genetic genealogists have done traditional, non-DNA-based genealogy research for decades. Other genealogists use genetic testing as but one research method to answer genealogical and historical questions, especially when the documentary record becomes thin or nonexistent. DNA tests, for example, are coupled with documentary evidence in surname studies. Y chromosomes (and surnames in most cultures) are inherited patrilineally. When the paper trail dries up, participants in surname projects can identify close relatives among other program participants through Y-chromosome test matches. As is discussed in chapter 2, a less commonly used but equally precise technology is mtDNA analysis, which clearly reveals genetic maternal lines of descent.

Extensive media coverage highlights the compelling nature of DNA tests for those who seek answers to their questions about identity. African Americans search for evidence to complement the documentary record disrupted by slavery that will link them to specific tribes or regions of Africa.[8] Those who self-identify as Euro-American or Caucasian work to trace their surnames to particular ancestors in the documentary record or to link their lineages back to particular villages in Europe. Many individuals who self-identify as primarily European American or African American look for molecular evidence of an Indian ancestor. I give talks all over the United States on the topics treated in this book, and often I am approached by audience members who say they have stories in their family of a Native American ancestor (most often Cherokee).[9] They wonder, can a DNA test prove their link to that tribe? There is anecdotal evidence of some parents and students looking for Native American genetic lineages that will enable them to check the "Native American" box on university and scholarship applications. In other cases, ousted tribal members have sought genetic evidence that they hope might get them back into the tribe.[10]

Media coverage further reveals a general lack of scientific knowledge on the part of consumers as they incorporate genetic-ancestry information into their understandings of personal identity, ethnicity, and race. Catherine Nash and Alondra Nelson have also described the use of genetic tests by genealogy researchers, finding that some users lack understanding of test limitations.[11] I regularly receive e-mail requests from such individuals across the United States who seek technical help with questions about how to apply genetic testing to their identity questions and family-tree research related to Native American ancestors. However, both Nelson and I have found that other genetic genealogists are propelled into a world of intensive and rigorous scientific research as they seek to deeply understand the technicalities behind their obsession. Indeed, whenever I receive those "out-of-the-blue" emails from individuals who are new to genetic genealogy, I always refer them to professional genetic genealogists and their blogs and books, and to the particular online community that I have studied. That community is the focus of the rest of this chapter.

The List: A "Post" Conversation

Alondra Nelson has described "contemporary genealogical research [as] a substantially technological pursuit." Like Nelson, I have found that doing social-science research on this topic has necessarily involved "virtual ethnography."[12] I observed participants mostly online in 2004 and 2005 in one genetic-genealogy community, through a Listserv that I refer to simply as "the list." During that time, I corresponded with a handful of individual "listers" and surveyed tens of thousands of posts archived since the list's formation, in 2000.

The list became an interesting and unique research site precisely because it was and remains populated by a wide array of researchers-cum-consumers. The list has its share of "newbies," those who are new to the science of genetic testing, although not usually to genealogy research. But many who are active and who post regularly appear to be very serious genealogical researchers who are also scientifically knowledgeable. Some are professionally trained scientists and engineers from multiple fields, and a few are professional genealogists.

List conversations often consist of brief, mostly technical exchanges. A typical exchange goes as follows: One lister posts a question to the list asking when two particular haplogroups diverged, and which was the original. A second lister responds that both are offshoots of yet another "superhaplogroup" found on a phylogenetic tree depicted in a particular recent academic article, which he cites. The second lister further explains that the tree doesn't indicate which haplogroup diverged first, but suggests that one could estimate the divergence by examining the accumulated variation within each haplogroup. The same lister notes that yet a third active lister has actually compiled a table showing estimates for ages of haplogroups that the third lister has drawn from a variety of academic publications. Yet a fourth lister, in a slightly snarky manner, disputes the third lister's data. There are seven additional posts to this thread, in which listers debate the technical merits and potential methodological problems with the table and the academic paper. This and similar discussions demonstrate that this group—at least those who are active—takes its science seriously. In this way, it is most likely not representative of the ways in which the broader public takes up the products of DNA-testing companies; these consumers are also meticulous researchers.

Genealogists self-subscribe to the list, often arriving in search of technical advice and conversation with other listers about which tests to take from which testing companies to answer their particular genealogy questions. But, of course, nothing is ever simply "technical." As feminist philosopher of science Sandra Harding explains, the technological, the social, and the political (not to mention the commercial!) are "always already inside each other."[13] I observed many complex and lengthy discussions on the list about the newest companies coming online and their technological wares, as well as sharing and discussion of the latest relevant scientific publications. Many of the most active listers spend considerable time honing their technical knowledge. When I was on the list in 2005, it was also known that testing companies were monitoring the list. At that time, two listers (not professionally trained genetic scientists) were heading their own DNA-testing companies. In a personal correspondence with the director of a U.K. company in August 2010, I was told that one of the two companies was no longer directed by the lister, and no longer existed in the United States. The director would reveal no information beyond that, except to say that the company was currently organized under different leadership in the United Kingdom. The second lister's company is still online, although, like many smaller companies, it is unclear how active it is in the market. Recently, as with the consolidation and failure of banks, we've seen some DNA-testing companies fail and others bought out by larger companies. Interestingly, the first lister in the typical conversation I have cited acknowledges that he has Native American ancestry, although he does not identify racially or ethnically as Native American. The second lister pointedly identifies as Native American, more specifically Cherokee.

In this chapter, I investigate through ethnographic, archival, and theoretical material several related questions: How do the genealogists on the list characterize DNA molecules and markers? Do they make them stand in for complicated understandings of life, humanity, history, and identity, in effect "fetishizing" them? How do listers talk about DNA in relation to race and ethnicity, in particular Native American race? Do they understand the roles of markers in identity in more complex ways? Do they understand Native American ancestry as an individual historical path, or as a quest for identity—as synonymous with race? How do listers reconcile Native American nations as a political category with their genealogical understandings of Native American ancestry? In short, for them, how do "genes" relate to concepts of "race," "tribe," and "nation"?

Method and Ethics: Observing, Participating, and Conversing

To investigate these questions, I reviewed online posts and related texts produced by individual list members. Although I was on-list daily and participated in list discussions from March 2005 through September 2005, I also read the majority of messages posted to the list each month throughout 2005, thus closely observing a twelve-month sample of posts.

In addition, I skimmed archived posts beginning in January 2000, when the list was founded, through the end of 2005, when I finished participant observation of the list. The number of messages posted per month during the first year of the list's existence (2000–2001) ranged from the teens to a few hundred. By 2005, the year that I was on-list, posts per month ranged between 1,200 and 2,700. Since I left the list, activity has declined considerably, likely because of shifts in how people do their online social networking. Looking back, activity on-list peaked in mid-2006, with more than 2,700 posts to the list that June. But beginning in 2008, posts began to decline noticeably, with 2008 having about 60 percent of the post activity of 2007. In the year in which I write, posts number from several hundred to nearly 1,000 per month. I asked Blaine Bettinger for his thoughts about the downturn in list activity. Although Bettinger offered no "empirical or objective data," in an e-mail to me on September 22, 2011, he had this to say:

> First, it's been a while since any big new product came out. I'd say that fall 2007 with 23andMe was the last big product in the genetic genealogy world (the 2010 FTDNA [Family Tree DNA] Family Finder test was a new option for test-takers, but wasn't a new type of test).
>
> Second, however, I believe that the single biggest change since the peak has been *how* people share and interact in the genetic genealogy world (and the world in general!). I think social media platforms such as Twitter, Facebook, and blogs have substantially replaced mailing lists as a means of querying and interacting with other genetic genealogists. . . . I've never liked the format of mailing lists, as I think they limit and/or inhibit interactions by being so user-unfriendly. I subscribe to the daily digest of the [list], and by the time I get notice of a conversation it's often already finished. In my opinion, social media platforms make sharing and interacting more user-friendly and efficient. I can't say for certain that this is a contributing factor, but I can say that I have many (200+) friends on facebook and twitter who are genealogists.

Indeed, Bettinger's Web site, *The Genetic Genealogist*, is a blog, news, and resource site that does much of the same work that the list does but in a more searchable format. Bettinger's site features a downloadable e-book, *I Have the Results of My Genetic Genealogy Test, Now What?*[14] Bettinger also refers readers to Smolenyak and Turner's primer for a more extensive education on the basic science of genetic genealogy. Through *The Genetic Genealogist*, Bettinger blogs about new genetic-ancestry testing products and software, posts notices of DNA-testing-company genetic-genealogy conferences and reviews of books on genetics and genealogy, and reposts news stories involving genetic-ancestry testing, such as "NFL Players Xavier Omon and Ogemdi Nwagbuo Confirmed as Half-Brothers."[15] Bettinger's blog, easily the most extensive of its kind, does much of the news sharing that the list did in recent years.[16] Like Ann Turner, Bettinger has a background that facilitates his doing genetic genealogy in a manner that is hard to describe as a hobby. In Bettinger's day job, he works as an intellectual-property and patent-law attorney. He also has a PhD in biochemistry and molecular biology.

Nevertheless, in the early to mid-2000s, the Listserv was the place to be. Indeed, in regard to the current politics of race and ethnicity in the early twenty-first-century United States, list conversations reveal a lot that is nuanced and current.

Because spending time on-list proved to be more valuable for my education than were initial e-mailed questionnaires, I did not pursue additional questionnaires or interviews. I destroyed the few responses I received, which were too few to provide a useful sample of listers. I also deleted related correspondences in favor of focusing on list posts. I did conduct one interview with a well-known genetic genealogist that proved instructive in how I might approach the field and individual genetic genealogists. In early 2005, I gave a talk on Native American race and genetics at Stanford, which Ann Turner attended. Turner is coauthor, with Megan Smolenyak Smolenyak, of a leading genetic-ancestry testing primer, *Trace Your Roots with DNA*. Turner later invited me to her home near Stanford, where I glimpsed, beyond her public and published interventions in the world of genetic genealogy, the depth of her commitment to the field. She had a large and well-equipped office and what looked like an extensive research library, including maps that I assume were of global haplogroup distributions. I knew that genealogy is a serious research endeavor and not simply a hobby for many of its practitioners. But nothing drove the point home like seeing a serious genealogist's work space. It differed not at all from that of a full professor (or, dare I say, an explorer), packed with decades of research and navigational materials.

Like other active genetic-genealogy listers and like genetic genealogist Blaine Bettinger, Turner has a scientific background that helps her closely follow the academic literature. Her undergraduate degree is in biology, and she has an MD as well. Judging from our conversation, the impressive array of resources filling her home office, and her frequent and helpful citations and syntheses on-list of current academic research, she is well studied in the latest genealogical thinking. Turner and other scientifically very literate genetic-genealogy listers regularly cite and discuss findings from recent scholarly publications from the fields of human-population genetics, biological anthropology, and related fields, including linguistics and archaeology. Sometimes their conversations about particular alleles and how they relate to particular ancestral lineages traced to especially European regions of the world are well above my head.

But Turner is not only scientifically literate. Her work demonstrates the kind of collaborative literacy I have come to expect of feminist scholarship, although Turner and I never talked about this. My interactions with her early on in my fieldwork served as a reminder that I should expect complexity and richness in the way that serious genealogists incorporate knowledge about genetics into their "obsession" with knowing their ancestries. Turner's professional prowess and her attention to the science and to the scientific development of other genealogists, her facilitation of productive social interactions between them (as evidenced in her interventions on-list), and her willingness to engage in conversation across difference raise the bar of conduct for all of us.

Taking my example from Ann Turner, I write the rest of this chapter as my next turn in a conversation with genetic genealogists whom I view as coconspirators in the compulsive process of research and knowledge production. In particular, and despite some essentialist language on-list (such as language that makes molecules stand in for humanity or life, or that relates markers to ethnic/racial groups in overly deterministic ways), I learned that genetic genealogists are clearly sometimes also anti–genetically essentialist. Some listers think in very complex ways about the relationships between genetic ancestry and identity claims, especially those individuals who have a decent grasp of the science and its limitations. In fact, despite their intense focus on genetic knowledge, they often speak with more complexity on the intersections of molecules and identity than does the press that covers these issues, or members of the public who show up at my talks and send me e-mail, or some academic scientists.

In the age of Google, and in a gesture toward anonymity, I anonymize lister commentaries and I do not quote posts word for word. Those who participated in the conversations or were on-list during 2005 may be able to put names to many of the positions described herein. That is unavoidable without gutting the content of posts.

Science, Religion, and Politics: On or Off Topic?

As in scientific venues generally, it is regularly asserted on-list that science and politics do not mix. The most heated debates target the perceived political corruption of what many feel should be a purely scientific topic and forum. In 2005, a lengthy thread brought the science-politics divide into full view and highlighted how that binary occludes a full accounting of what counts as and who wields "politics." The thread had to do with U.S. and New Zealand indigenous opposition to human genome diversity research. The Indigenous Peoples Council on Biocolonialism (IPCB) had issued a press release that challenged the colonial nature of DNA research and that asserted indigenous peoples' property rights to their DNA and to the knowledge produced of it. Listers essentially took up only one of IPCB's critiques, that indigenes had a right to not have their cultures undermined by genome research. The indigenous critics were variously characterized by several listers as clinging to myths, as not really believing in the myths but using them to political ends, as dependent on government welfare and outside the mainstream, and/or as self-appointed spokespeople. Several listers were more generous in defending indigenous peoples' rights to believe in their myths, untrue as they might be. The lengthy discussion centered on the conflict between traditional and genetic knowledges and the greater veracity of science, but with the caveat made by at least half of those involved in the thread that cultural differences had to be respected.

A key element of IPCB's critique was that scientists historically have taken indigenous DNA without proper informed consent. Some had even lied (or lied by omission) to their subjects about the type of research that would occur on their samples, such as in the Havasupai and Nuu-chah-nulth cases, in which indigenous peoples in Arizona and British Columbia, Canada, respectively, were denied their rights to fully informed consent when their samples were used for research on ancient human migrations and population history to which they objected. In both cases, researchers treated indigenous samples as their property to do with what they wished.

IPCB spokespeople argued that whereas scientists would benefit from genetic research, it was dubious that indigenous peoples would benefit. Thus, although cultural differences were certainly at play, IPCB was really focusing on the imbalance of power that exists between genome scientists and their indigenous subjects. Listers ignored IPCB's arguably weightier claims—that indigenous people had a right to full disclosure and control of the uses to which their DNA was put, and that they had a right to expect tangible benefits from research, to be not simply the biological ground upon which all personkind might ultimately benefit. Several listers reminded their colleagues that some indigenous people would be interested in such research (IPCB doesn't speak for all). They hoped that others would eventually see the light and participate in research that would ultimately benefit everyone. If, however, some groups chose to remain in the past, they would be left behind to their own detriment. The multicultural-difference argument and the ideal of individual choice reigned on-list. Rather than acknowledging the power relationships set up in colonial history that still condition the way genome science gets done, all but a few listers held that indigenous peoples were on the same equal footing with scientists, genetic genealogists, and other "rational" consumers of genome science. Indigenous peoples, like everyone else, had a "choice" about whether or not to participate and thus benefit. Yet IPCB's critique was precisely that indigenous peoples have not had a choice and have not benefited. Within the science/politics binary, listers could not see that scientific subjects and practices have all along been simultaneously political subjects and acts, something IPCB understood all too well.

Unless they die a natural and timely death, posts deemed to be about politics and not science are eventually declared "off topic" by annoyed listers and ended by the list administrator, who declares an "END OF THREAD." The list administrator sometimes suggests alternative Listservs where religion or politics (even as they relate to DNA) can be debated. Staying "on topic" is a goal enforced fairly tightly on this list. It is for discussion of DNA- and genealogy-related topics narrowly construed. That there should be no religion or politics is generally agreed upon, at least by those who are most vocal.

What, in addition to the IPCB critique, constitutes religion and politics for listers? "Political" discussions often begin with a breaking news story, whereas others begin with a technical discussion—for example, how to interpret and resolve somewhat conflicting ancestry-DNA test results. One such thread developed into a non-DNA-related discussion on the politics of illegal immigration at the U.S.–Mexico border. It is rare, but I saw a racial epithet used in that exchange. The act elicited quick condemnation from other listers. It was declared inappropriate and off topic.

Posts that delve too deeply into religious talk are also viewed as inappropriate. Threads promoting purportedly American Indian religious perspectives have been quashed on-list. In one thread, a lister posed a question: given that some on-list had DNA that they knew reflected Native American ancestry, how might tribal nations handle the DNAPrint test that has trouble reliably distinguishing Native American from East Asian ancestry? A second lister, who identified as Native American, specifically as descending from a Cherokee chief (as well as from populations in the British Isles), replied essentially that the lister's post was a nonquestion. He clarified that tribal nations don't accept genetic-ancestry testing evidence for enrollment, which is generally true. In addition, he argued that tribes do not see it as useful, because they don't acknowledge as valid this particular "white man's technology" produced of a particular "dogma" about the first settlement of the Americas. Yet a third lister, a woman, shot back that the Cherokee-identified lister should clarify his use of "dogma" in relation to the peopling of the Americas. He directed her to read Vine Deloria for clarification of Native American critiques of scientists' New World migration narratives. She replied that if the Cherokee-identified lister was talking about "mythology" and "religion"—and she respected his right to believe in them—they had no place on-list.

The reader might expect that I sympathized with the position of the Cherokee-identified lister. I found myself being as annoyed with him as his correspondent, but for perhaps not all of the same reasons. Ironically, he is the Cherokee-identified person mentioned earlier who operates a DNA-testing company. The lister's analysis in this exchange about why tribes might reject DNA testing was thin and incongruous with his authoritative tone, which does not come across in my summary of the conversation. There is much more at play with tribal positions for or against genomic research and DNA tests than a religion-versus-science divide, as suggested by IPCB's critique of traditional power relations between researcher and subject in human genome diversity research. The Cherokee-identified lister represented Native American tribal stances on such things in an overly generalized manner in order to gain the moral high ground. His position perpetuated a stereotype about Native American tribes' actual or potential relationships to genome science, and to science and technology more generally. Aside from being not wholly accurate, such characterizations potentially limit the range of possible responses to science and technology that tribes are able to legitimate.

Other conflicts over religion on-list involve listers who bring religious doctrines or narratives as sources of evidence or historical truths into DNA research, for example, to reject or question genomic evidence that contradicts religious history. One lister posted favorable comments of Mormon scholars who attempt to use genetics to support church views of creation and the settlement of the Americas. He was roundly criticized and his views were declared irrelevant to the list. The majority of disagreements over religious perspectives involve Christian religious doctrine of one form or another, which is not surprising given that Christianity, broadly speaking, is the dominant religion in the United States and this is a list dominated by Americans.

It is interesting to me that Native American and Christian perspectives that are critical of genome knowledge are seen on-list to fall on the same side of a religion-versus-science divide. Unlike Christian traditions, Native American origin narratives are generally missing the will to convert and so are without inherent intolerance for other ontologies. But most important, Native American resistance to genome research also always focuses on who has the power to research whom and how, and who has the power to make policy that affects Native American lives. Again, this incisive critique—as we saw with the IPCB thread—was largely missed by the genetic genealogists, just as it is by scientists who make the same false comparison between Christian creationists and tribal creation narratives, notions of the sacred, and political resistance to being objects of research.

Another incident in which science was seen as getting watered down in its rigor by politics involved one lister arguing that some Native Americans are, in effect, really Europeans. Referring to the least common Native American mtDNA haplogroup, X, the lister argues for the hypothesis that "X people" traveled along an ice floe across the Atlantic during the Last Glacial Maximum from Europe. These became "Clovis people," referring to late Pleistocene North American peoples credited with producing the unique "Clovis point." Thus, the lister argues, Native

Americans are actually Europeans. He is not the only one to make such a claim. A science reporter does the same thing in a 2006 LiveScience .com article, "First Americans May Have Been European."[17] The lister refers to the Solutrean hypothesis (though he does not name it), advanced in a controversial 2004 paper published in *World Archaeology*.[18]

The lister somewhat aggressively anticipates the fall of the "old theory," presumably the Bering Strait account of human migration to the Americas. (An interesting point of background on this lister is that in other unrelated threads he expressed views that assumed the veracity of Christian religious doctrine and was disparaged by active listers for bringing such views onto the list.) At this lister's overly authoritative dismissal of a long-established scientific theory, a second lister suddenly could not hold back and jumped into the conversation with a tone akin to "I can't take it! This is ridiculous!" With a few words, in all capital letters, he clarified that the Solutrean hypothesis was still, in fact, a controversial new theory and that it was actually not yet established that the X haplogroup came to the Americas from Europe. He then cited at length some of the problems with the scientific evidence for the hypothesis. The first lister responded with two separate, brief comments that essentially denigrated academic science (I think) in favor of his Family Tree DNA test results, which seemed to say something different to him. Perhaps the scientifically more well read lister thought the remarks confusing, too, as he refrained from a second response. The conversation petered out.

The Politics of Studying Scientific Subjects

Politics on-list are not limited to discussions that listers readily recognize as entering the realm of the political, such as those that let slip a racial epithet or that overly corrupt a scientific hypothesis with religious or other "cultural" doctrines. Much of the politics at play on-list are not acknowledged by listers as such. When my research came to their attention, in March 2005, it generated controversy within the list's recurring science-versus-politics debate. In early March, I asked on-list for volunteers who had attempted Native American DNA analysis to complete a short questionnaire. Fewer than ten agreed, and those individuals did so quickly. Shortly thereafter, one lister who had agreed to be surveyed came across a notice for my March 7, 2005, talk at the Stanford

University Humanities Center, "'Native American DNA' and the Search for Origins: Risks for Tribes," on the center Web site. I was one among two-dozen-plus speakers participating in a two-year series, "Revisiting Race and Ethnicity in the Context of Emerging Genetic Research."[19] Upon seeing the context of my talk, the lister informed me that he was no longer interested in being interviewed.

I corresponded with the lister both on- and off-list about his dismay with my research focus and his subsequent refusal to participate. However, I draw on his posts to the list only to describe his hesitance. He was actively critical of social-science theories and approaches to understanding the field of genetic-ancestry testing, which became even clearer in a subsequent on-list debate over the phrase "gene fetishism," which I'll discuss shortly. His primary reason for declining to complete a questionnaire, stated on-list, was that his own commercial ventures in genetic genealogy might well be a target of my critique. Why should he offer up himself as a subject in a research endeavor that could undermine both his own research and his commercial activities? Fair enough. My research could undermine his way of life. A longtime genealogist, the lister had recently become a director of a DNA-testing company. (He is the second lister mentioned earlier, who previously directed a U.S.-based genetic-ancestry-testing company.) He was particularly disturbed by my focus in the Stanford talk on the "essentializing" representational practices of some DNA-testing companies. Ironically, his company Web site displayed care with language and imagery, avoiding the kind of sensationalist exaggeration of the science in which some companies indulge. After viewing his company site, I was encouraged that companies could in fact resist essentializing representations that underplay the complexity of social life and genomics as they relate to human identities. His company's site lacked the hyperbole of the companies I highlight in chapter 2.

I later read posts that indicated that this lister chose company Web site language in ways that foreground the scientific imprecision in ancestry DNA as related to race in order to avoid critique by the "politically correct." I was certainly not the only critic being referenced by the lister; the entire Stanford lecture series sparked debate on-list. Such "PC" individuals, the lister argued, might interpret his company's work in a "political" manner were the company to use "the 'r' word"—*race*, it can only be presumed—in relation to its products. His company therefore chose an alternative term, "genome diversity." But it chose this term more for the fact that it avoided "race" than for its greater scientific precision. At that point in the thread, the lister fell into an ad hominem attack. He did not debate in substantive terms the critiques of the "leftists" he saw himself as opposing. Rather, he accused critics of being career-ambitious, self-proclaimed mediators of morality with "revisionist" historical agendas who were consumed with tearing down what they perceived to be "fascist" genomic tools that categorize human beings. His assault on the scholarship in my area was an especially educational moment. I should note that the lister could be patronizing, but he never attacked me personally. In fact, he indicated on-list that he sympathized with my desire to get a PhD and the challenges that involved. (He had emphasized his own PhD in list exchanges before.) He also speculated about my position (at the time) as a graduate student who was most likely powerless and should not be held completely responsible for the language and approaches I used.

Disciplinary Cultures and Conflicts

This same lister also noted that he felt patronized—and so did others of his compatriots—by nongenetic scientists who study race and genomics. He and some others felt that academics in my area of study believed that genetic genealogists were all unreflective about the intersections of genealogy, genomics, and race.

To the contrary, the lister's analyses of the relationships between race, genomics, and identity were nuanced, especially his understanding of the limitations of genetic-ancestry testing for Native American and First Nations citizenship. I read in archived threads his discussion of the misinformed expectations of some genetic-ancestry test takers who might hope to gain U.S. tribal or Canadian First Nations enrollment based on test results. He noted the absence of genetic markers that would indicate tribal-specific ancestry, which is what tribal governments are interested in and which they gauge through paper documentation of enrolled parents or grandparents. The lister acknowledged that although many individuals who seek DNA tests probably do have Native American ancestors, the best evidence that the genetic-ancestry-testing industry has to offer is not evidence that tribal-enrollment regulations consider.

The lister even explained that his company did not endorse using its services to pursue tribal affiliation, especially as sole evidence. Nor would scientists at his company testify in court on behalf of customers' tribal-enrollment claims, and customers must sign a document noting that they accepted such conditions. The lister went on to explain to the list the complicated and sometimes contradictory ways in which Native American identities are formed. He noted the intersections of descent from Native ancestors, self-definition, and a Native American identity ascribed or withheld by others based on physical appearance, reservation residence, knowledge of tribal histories, and social life. This lister emphasized that a socially reckoned Native American identity does not always line up with tribal-enrollment regulations. His take on the situation (like much of the discussion on-list) was a breath of fresh air in a broader world in which genetics and identity seem to get conflated at every turn.

As I mentioned earlier, I tried to be careful with my words and respectful in my engagements, for several reasons. First and foremost, the list was populated by many retired individuals—my elders. Being raised by old women, I learned that one does not speak to a seventy-year-old like one speaks to one's thirty-six-year-old peer (my age at the time). Second, the list is publicly accessible; what one puts out there, all of cyberworld can see. I tried to balance respect for people whom I did not know personally but who were in many cases much older than me with a substantive, intellectual engagement. And many listers, because they are researchers, take some pleasure in argument. I was genuinely pleased with the complex conversations I was able to have with some of them. I appreciated the time they took to engage with me. I also appreciated that even when we disagreed, they pushed my thinking. I learned in conversation with listers. Did this particular lister view me as patronizing him, or was he referring to a patronizing tone he regarded as generally used in the social studies of science?

Another lister, a woman, also clearly felt patronized. Like the male lister, she also wrote complex and interesting posts on the intersections (or not) of genomics and identity. She would rein in other listers' sloppy scientific assessments or overly genetically determined characterizations of what constitutes, for example, a Native American tribal member. It quickly became clear that comments about patronizing academics had a lot to do with differences in language. At one point during a thread about "gene fetishism" prompted by the presence of the term in my talk at Stanford and a Donna Haraway reading I had cited, this second lister berated so-called postmodern academic discourse for its elitist and inaccessible language. Haraway is famously difficult for the uninitiated to read. In her chapter "Gene" from *Modest_Witness@Second_Millennium.FemaleMan©_Meets_OncoMouse™* (1997), Haraway explains "gene fetishism" in relation to Marx's "commodity fetishism":

> Commodity fetishism was defined so that only humans were the real actors, whose social relationality was obscured in the reified commodity form. But "corporeal fetishism," or more specifically gene fetishism, is about mistaking *heterogeneous* relationality for a fixed, seemingly objective thing. . . . Gene fetishism . . . denies the ongoing action and work that it takes to sustain technoscientific material-semiotic bodies in the world. The gene as fetish is a phantom object, like and unlike the commodity. Gene fetishism involves "forgetting" that bodies are nodes in webs of integrations, forgetting the tropic quality of all knowledge claims.[20]

The second lister saw language such as "gene fetishism" as simply catching up and intoxicating individuals, rendering their writing completely inaccessible to those outside the field. She went on to describe writers who engage in such discourse as silly, self-indulgent, and caught up in a diminutive subculture.

Charges of elitist language were paradoxical coming from genetic genealogists who tolerate much inaccessible, technical genetics terminology on-list. Many of their exchanges are detailed, focused, and sometimes even passionate discussions threaded through with insider abbreviations, technical terms, and references to journal articles with very small and specific audiences. Active listers clearly enjoy wielding, with varying degrees of fluency, the scientific languages they study. Indeed, they are caught up—entranced even—by the discursive-material cultures of human-population genetics and biological anthropology, as those areas of research inform the genetic-ancestry-testing industry. Listers wrote with great interest of the most recent findings of scientists whose work they followed. They wrote glowingly of correspondences with scientists and scientists' presentations at international genealogy conferences. Listers were also critical of the industry and its limitations. Such listers had a command of scientific concepts and technical language that comes with diligent study and is a source of both utility and pleasure.

But, like too many academics, listers seemed to have adopted disciplinary chauvinism on behalf of human-population genetics. The "gene fetishism" thread was lively and involved multiple active, technically knowledgeable listers. With one memorable exception, those who commented at length on the gene-fetishism topic were sure that social-scientific specialty language was simply without intellectual merit. Its use, therefore, was seen as showing a lack of moral integrity. Why was it acceptable for genomics in and of itself to be complex, but theories about the interplay *between* several additional complicated realms—genomics and political and cultural economies—must be easy for everyone to understand?

This entire book is a critique of the idea that science and society are separate domains. Part of viewing them as separate entails the idea that "hard" science is difficult, and society or "soft" science is easy. In addition, because we labor under that other presumed binary, the academy versus the "real world," there is a tendency to believe that academic discourse, especially when it is not hard science, is difficult only because it means to obscure and exclude. It is "elitist" and "out of touch" because it is not actually useful. I counter that we need precise languages to talk about precise ideas that have derived from specific histories of work, from the development of theories and methods. Those of us in the hard sciences need them. Those of us in the social sciences and humanities need them. I regularly encounter this type of disciplinary chauvinism among academic scientists. Like listers, they often refer to social theory as "jargon," as if they should readily understand what it has taken me and other social scientists and humanists years to master. I do not assume that I should readily grasp all of the language used and data introduced in a technical presentation about the genome diversity of oak-tree populations in Northern California. Perhaps they have the false impression that I do this work because it is easier than what they do. The arrogance of that position disheartens me profoundly.

The majority of listers did not weigh in on the debate; perhaps ten did. I had several supporters on-list, especially individuals who evidenced concern about the effects of genetic ideas and DNA testing on tribal enrollment and Native American identity. One lister in particular comes to mind as alternately criticizing and supporting my ideas. He started out being a vocal detractor of my proposed work. Over the course of my time on-list, he eventually agreed with me that the DNAPrint test is problematic in terms of both its technical and its racial-type assertions. But he disagreed that one of the "great" male scientists whom we discussed on-list (James Watson) is actually guilty of gene fetishism. After my time on-list, James Watson espoused another arguably genetic-fetishist view. He said in a widely and negatively publicized 2007 interview with the *Times* of London that he was "inherently gloomy about the prospect of Africa," because "all our social policies are based on the fact that their intelligence is the same as ours—whereas all the testing says not really." Watson also thought genes responsible for human intelligence could be found within a decade.[21] I wonder whether my former listmates might now reconsider their assessment of Watson.

One other lister charged scholars who ponder gene fetishism with practicing "pseudoscience." Given her lack of substantive analytical engagement with the term "gene fetishism," her charge of pseudoscience might be interpreted as constituting any research that produces conclusions with which she disagrees. One can interpret the pseudoscience allegation as rooted in a belief that any scientific or disciplinary approach that falls short on traditional notions of objectivity—the disembodied "view from nowhere"—constitutes so-called pseudoscience. By default, "real science" is placeless, genderless, classless, and raceless in that it should be replicable by anyone regardless of those identifications. Within such a view, analyses that call attention to the cultural and political economies inside which the sciences get practiced and which shape research questions, methods, and data interpretation are not permissible. Studies such as mine attempt to make explicit what is already implicit in scientific practice. Listers "know" that there is truth in this science, and they cannot reconcile that with "politics." As a result, those scholars who call attention to the existing politics of genomics and anthropology are the pseudoscientists—the ones who impose unwarranted political interpretations.

I encounter this response regularly, including from scientists. A representative incident occurred at the American Society for Human Genetics (ASHG) annual meeting in 2008. ASHG is heavily attended by academic and industry-research scientists and clinicians. On an "ethics" panel (a type of panel that, revealingly, is always separate from the scientific panels), I gave a talk that pointed out old-fashioned understandings of race that inform research projects on human migrations and human genome diversity. At the conclusion of my talk, an audience member, a middle-aged woman, stood up. She was so angry that she shook, and she charged me and our panel with "imposing" politics onto genetics. It is difficult for those who accept the science-versus-society binary to consider the situatedness of science and scientists, of research questions, methods, and interpretations. The partiality of scientists' perspectives conditioned by where they stand within a particular historical moment and context would seem to discredit the entire project of science. In that view, "politics" are seen as presenting barriers to getting good evidence and effective analysis instead of as always already present, as something that must be accounted for if we are to get better evidence and do more effective analysis. This is why strengthening our notion of "objectivity" to account for multiple, diverse standpoints, or "situated knowledges" as Sandra Harding and Donna Haraway emphasize, is so important.

An important aspect of lister identities is their identification with scientific rationality. Key to their defense of their own research practices was the defense of "Science" and scientific ways of thinking. Critical analysis of genomic science, especially the world of genetic-ancestry testing, was seen as "denigrating" genetic genealogy and attacking practitioners' character and judgment.[22] The gene-fetishism discussion back in 2005 had already made clear to me listers' defense of (genetic) science not only as a way of life but, for some, as a superior way of life, especially as compared to some of the misguided endeavors of social science, such as applying Marx's insights to an analysis of the role of molecules and genes in twenty-first-century culture. The congruence between lister responses and the scientist at the ASHG meeting is striking. Both are scientific subjects.

Scientific Subjects Also Resist Being Researched

My encounters with listers were paradoxical on another level. The tables seemed suddenly turned from Vine Deloria's critique of white-on-Indian anthropology, when he berated "anthros" for doing "PURE RESEARCH" on American Indians—research that is "absolutely devoid of useful application," "an abstraction of scholarly suspicions concerning some obscure theory."[23] The paradox was clear. Here was I, a Native American woman from a somewhat class-disadvantaged background, berated for engaging in obscure theory and what was basically anthropological research on white folks, many of whom were clearly financially able, as evidenced by their ability to purchase DNA tests costing hundreds of dollars, their computer access, their travel to genetic-genealogy conferences, and their time to be online much of every day.

Listers foreground social-scientific academic discourse as the source of condescension, and language is a serious challenge for conversing together across disciplines and life projects. We could view exchanges between listers and me and other race and genetics scholars as simply reflecting the challenges of "cross-cultural" communication and understanding. In fact, that assessment of why communication is difficult is typical when scientists find themselves challenged to work effectively and collegially with Native Americans and others who resist their research agendas. In order to combat these "cultural" difficulties, scientists and funders regularly call for more "public understanding of science," or PUS, an unfortunate acronym indeed. But ironically, like Native Americans, who sometimes resist being researched, I read listers as registering their objections to being researched, period. It is not simply that social scientists are wrong (and not all listers agreed that we are wrong), but how dare we make listers and their field objects of research? Paradoxically, we social scientists are seen as simultaneously arrogant and elitist in our use of language and also inadequate and, unlike genome scientists, not worthy of studying them. Our projects are not worthy. Our knowledge is "ivory tower," out-of-touch pseudoscience.

Active listers clearly spent considerable time and effort in doing genealogy and genetic genealogy. They participated deeply and frequently in technical discussions on-list and encouraged new genetic genealogists, "newbies," as they attempted to grapple with the science and its networks. They evidenced a deep investment in the community by their consistent presence there and their considerable intellectual work in service to furthering the broader body of knowledge. Foregrounding my Native American subjectivity, from where I stand, I saw listers as members of a dynamic community with considerable expertise and a growing body of rich and interesting knowledge. Yet listers and their field were put in the position of serving as objects of my study. They willingly offered themselves up as objects of study, in a sense, when they purchased DNA-test kits. They were agreeable to being subjected to study by genome scientists. Genome scientists produce knowledge that listers need in order to do their own genealogical research. Genomics facilitate genetic genealogists' ability to build their "family trees" and their deep historical narratives. Genomics and biological-anthropological knowledges also give listers pleasure. It is a mutually beneficial relationship. But how do listers benefit from the knowledge that humanists and social scientists such as me produce? From where they stand, my work and that of other critical social scientists risks impeding their work and their way of life. This is the same reason that genetic genealogists tend to oppose regulation of the direct-to-consumer genetic-testing industry.[24] Lo and behold, the genetic genealogists felt like Indians!

A Note on Gender

I did not make gender a target of my analysis, but it seems important to address, especially in relation to the previous discussion of scientific discourse and subjectivity. When conversations became, as one lister described, "vitriolic" or "bombastic," I noticed a correlation between gender and communication style. During the year I was on-list, when vitriol did effectively shut down conversation and an "END OF THREAD" post from the list moderator quickly followed, it was without exception the work of a male lister. I frequently read female lister posts that could be characterized as annoyed or sarcastic (some of them directed at me and my colleagues), but I saw only one that qualified as shutting down conversation. However, that particular female lister did not shut down conversation with abusive or slandering language. Rather, she took up the position of victim. She was fairly active on-list, but in this case she was in conversation with one of the more technically competent male listers, a person who had been called out by other listers before for his direct (and sometimes offensive) communication style. Indeed, I often witnessed communiqués of his that did not mince words when he disagreed with another lister's interpretation of scientific literature. In this case, the female lister viewed his posts as a personal attack. As I read it, the male lister's posts did not reveal the level of hostility she perceived, and the female list administrator at the time indicated as much. It is interesting (or perhaps not) that the female lister so quickly took up the subjectivity of victim in relation to the male and scientifically more competent lister. By contrast, several active and technically competent women on-list come immediately to mind as consistently not shutting down dialogue even when they had much at stake, for example, when they were engaged in defending science from politics.

Unfortunately, data was not available on the number of male listers in relation to the number of female listers, nor on the ratio of male-to-female posts on-list. During the time that I was on-list daily, I noticed a healthy level of participation by women among active listers. But as I describe noticeable gender differences in communication style, I wonder what it would take for me to notice low participation by women. Would fewer than 50 percent of posts by women grab my attention? Would it take fewer than 25 percent, or even fewer than that? Because I noticed that women sometimes communicated differently on-list and because science has long been too much the domain of men, I might have paid unusual attention to posts by women and overrepresented their numbers in my thinking.

In October 2005, I sent an e-mail to the new list administrator, Jim Bullock, asking him about numbers of women versus numbers of men on-list. He indicated that his list of subscribers provided only e-mail handles, many of which did not indicate full names and thus did not indicate gender. But he offered me two personal observations about gender in the genetic-genealogy community. When he attended Family Tree DNA's First International Conference on Genetic Genealogy, in 2004 in Houston, which invited only surname-project administrators, he noted that of 170 participants, women seemed well represented. Meaningful participation by women both at the conference and on-list is interesting. Surname projects and Y-chromosome tests make up the bulk of genetic-genealogy work, and women, of course, do not have Y chromosomes to analyze. Yet Bullock noted that women are "often the primary family genealogists, so they may often be the better informed family members when it comes to DNA testing." Perhaps I am not wrong, then, in remembering women as holding their own in this scientific domain.

The Listers "Do" Race

When I stumbled onto the community that is this list, I knew next to nothing about the world of (genetic) genealogy research. I was surprised by the segregation I found there, although I should not have been. I quickly figured out that the list comprised largely self-identified whites and a few "Europeans," but mostly "Americans."[25] Even without a survey of the racial or ethnic backgrounds of listers—such numbers were unavailable—the race identities of listers who posted usually became clear. They not infrequently declared on-list their racial or ethnic identifications as "white," "WASP," "Caucasian," or the like. There were fewer declarations of Jewish identity (whether "white" or not), although there was at least one very active Jewish-identified lister. Thus, discussions of Jewish-related DNA markers were not infrequent. The only other explicit race-ethnicity identification I encountered on-list was Native American of one type or another.

A second indication of the overwhelmingly white racial makeup of the list can be inferred from the range of particular "surname projects" and associated locations that listers mentioned. There are far too many to list, but those surnames were predominantly of European lineage. Surname projects are crucial to the work of genetic genealogists. Many got their start in DNA research because of them, finding their "DNA cousins" in Europe and in settler countries such Canada and Australia.

A third indication of the overwhelmingly white racial makeup of the list was the race-ethnicity categories invoked by listers as they did their research. Within the nearly twenty-five thousand messages that I scanned in 2005, I counted references to 282 different "biogeographical," race, ethnicity, nationality, and/or tribe categories. Some were referenced only once, others in hundreds of posts. How did I count? Sometimes two such references appeared in one subject line, such as "Flemish influence in Scotland" or "First Nations . . . and Being British." In such instances, I counted both citations. I use the shorthand category of "race-ethnicity" to refer to such citations. Some listers might reject this category. They might instead prefer "biogeography" or "ancestry," for example, as an overarching label for their references to "European," "Cherokee," or "Scots-Irish." However, I contend that race is always (also) at stake even when listers take pains to articulate categories that they hope can be distinguished from concepts of race that might seem outmoded or contentious.

Estimating conservatively for 2005, roughly 70 percent of lister race-ethnicity references pointed to what we call today Europe, for example, "Albanian," "Deutsch," "Eastern Europe," "Hungarian Scots," "SW England," and "Swedish." I also include "Viking," because it is strongly associated with present-day Scandinavia. But that 70 percent does not include references that might point to ancestry in Europe that is considered racially mixed (such as "Hispanic") or ancestry that is highly diasporic (such as "Jewish") or ancestry speculated on-list to be, in part, outside of Europe (such as Irish "travelers"). Nor does it include references to Central Asian or Middle Eastern regions and countries, whose peoples are often arguably difficult to disentangle from European populations.

Interestingly, I encountered a problem in my research similar to that which genome scientists encounter: how did I decide which populations to put into the "European" category in order to get at something ultimately termed "white"? Genome scientists exclude admixed samples. I, too, excluded subjects that would trouble dividing lines. In addition to the "European" references, posts related to U.S. European-descent or "white" populations (such as to "Mayflower Pilgrim" lineages or to the "Pennsylvania Deutsch") numbered approximately 3.8 percent of race-ethnicity posts during 2005. I incorporate these into the 70 percent of "Europe"-related posts.

In counting references to Europe, however, I exclude many references to particular Y-chromosome and mtDNA haplogroups and types designated by letters and numbers that are traceable to particular regions of the world. Not surprisingly, I find that haplogroup/type references correspond in number to explicitly noted race-ethnicity references. The vast majority refer to lineages found at highest frequency in Europe. But if I were to count such references as race-ethnicity posts, I would have to judge how to count, for example, a haplogroup that has been found at a slightly higher frequency in "European" populations versus "Middle Eastern" populations. This would compel me to sort through multiple lister classifications, as those are explicitly or implicitly noted in post texts, and then make decisions about how to categorize beyond listers' self-categorizations.

I tried to be conservative in constructing categories, replicating them exactly as I found them in the subject lines of posts. That method also made for easier searching and tallying of race-ethnicity references. I did combine, from across months, categories that are essentially the same but that were typed differently (such as "African-American" and "African American"), correcting typos and misspellings. However, I resisted collapsing, for example, "Central Asian" and "East Asian" into the broader category of "Asian" in order to replicate, as faithfully as possible, lister categories. "Central Asian" could allude to "Asian," but it might also come up in a series of posts about "Eastern European" lineage. Would I count the "Central Asian" references, then, as part of "Asian" or "European"?

In my counting of potential race-ethnicity references, I also excluded citations that could be taken as related to biogeography, nationality, or race-ethnicity, including, for example, "Norwegian Earthquake," "Tsunami in Scotland," "Scottish Declaration of Independence," "Norman Conquest," "Wooly Mammoths in Florida," and "The Silk Road." Such references might have explicitly discussed human-DNA research in the aforementioned parts of the world, or they might not. Some subject-line references to "Native Americans" actually did not involve genealogy research; they referred to the ethnic studies scholar Ward Churchill and the controversy surrounding charges that he fraudulently claimed to be American Indian. When links between subject-line references and genetic content were in doubt, I investigated the text of posts closely.

African (American) Ancestry (Not) On-List

I found not a single declaration of African, African American, or black identity on this list during my time on-line. Self-identified white listers never sought recent ancestry in Africa. As a result, only about 2 percent of posts to this list during 2005 had to do with African ancestry. Because longer-ago so-called African ancestry can be claimed by all humans, nearly one-half of African-related posts had to do with African lineages in antiquity (with subject lines such as "Out of Africa" or Africa as the source of "human origins"). Only fourteen posts during 2005 mentioned the category "African American" or "black," and some of those commented on celebrity genealogy projects such as one conducted by Harvard's Henry Louis "Skip" Gates Jr. A single subject line during 2005 mentioned slavery, a central topic that African American genealogists grapple with in their research.

It would seem on-list that to self-identify as racially white does not correlate with having or suspecting recent African ancestry in one's genome, which is unsurprising given the concept of hypodescent (ascribing offspring of a "mixed" mating to the parental racial group that is or has been considered inferior) that structures our dominant black–white race binary in the United States.

Or did listers simply go to other venues for such conversation and assistance? Were they, for example, frequenting the Listserv associated with African American geneticist Rick Kittles's company, African Ancestry, which requires a person to purchase one of his DNA tests to join? It doesn't seem so. In her research on African American genetic genealogy and on Kittles's work, including the Listserv associated with his company, Alondra Nelson never mentions white (or Native American or any other racially identified) genealogists. She regularly uses the terms "black" and "African American" to describe her research subjects without clarifying how their identification as black was figured prior to engaging in genetic genealogy. There is a lot of thoughtful consideration post-DNA-testing by Nelson's subjects over whether they are really or technically "Ghanaian," "Hottentot," or the like based on their documentary and genetic research, which sometimes conflict. But we are simply presented with their panracial designations as "black," "African American," or "black British." Nelson also noted in personal correspondence that although she doesn't recall anyone "nonblack" on her list (and she wasn't looking for them), "there were folks who identified as bi- or multi-racial."[26]

In Nelson's study population, individuals' identity as "black," "African American," or "bi- or multi-racial" is determined in whole or in part by having recent "African ancestry." Fair enough. Yet many of the genealogists she studies are also clear that they have European and/or Native American ancestry as well. The history in the United States of racial assignment by hypodescent and the black–white binary that has resulted in U.S race relations has made it difficult or impossible in practice to have one's race legitimated as white when one has recent African ancestry. On the contrary, it is the norm to identify as black even while clearly also having recent European or Native American ancestry.

De facto segregated Listservs reflect these racial histories. (Note that for Nelson's Listserv it was also possible to identify as "mixed-race" and African American.)[27] On the list I frequented, "mixed-race" did not arise as a frequent category of self-identification, although some individuals identified as being descended from both European and Native American populations.

Native American Ancestry On-List

I came to the list expecting some conversation about Native American ancestry. Genetic testing can be a logical alternative for those who have trouble legitimating their Native American ancestry. Given the very low number of self-identified Native Americans within the broader U.S. population—0.87 percent or 1.53 percent, depending on how we are counted, as Native American only or as Native American mixed-race (the numbers are even lower if we count the tribally enrolled)[28]—and given the otherwise racially homogeneous demographics on-list, I didn't expect the magnitude of Native American–ancestor searching that I found. For 2005, I conservatively tally 8 percent of race-ethnicity listings as having to do with Native American populations and ancestry. I do not count "East Asian" references, for example, that often have to do with research on Native American ancestry (because of the difficulty for DNAPrint's AncestrybyDNA™ test to distinguish Native American from East Asian lineages). Given my knowledge of Indian Country and after reading relevant posts, the subject lines of 2005 posts that I do count as "Native American" include "Alaskan caveman DNA," "Amerindians," "American Indians," "Black Indians and Five Civilized Tribes," "cherokeendnDNA," "Creek," "First Americans," "First Nations," "Indians," and "Lumbee." The vast majority of references I have seen throughout list archives (maintained since 2000) refer to panracial designations, such as "Amerindian." In all years (2000–2005), I have encountered many fewer specific "tribal" ancestries in the subject lines of posts. I also group "Melungeon" ancestry under the Native American umbrella. It is an ancestry consistently discussed on-list. "Melungeon" refers to a multiethnic/multiracial people who are centered in Appalachia and who claim Native American ancestry, among other claims. In list conversations about Melungeon history and identity, Native American ancestry was front and center. Not surprisingly, the specific tribal/ethnic groups mentioned here are those that acknowledge histories of extensive "admixture," as geneticists would call it, with non-Native peoples. This relates to anthropologist Circe Sturm's observations that the Cherokee are targeted by "race shifters"—whites seeking a Native American identity—because "Cherokeeness seems open to whiteness in ways that Navajoness [for example] does not. . . . Cherokeeness is an ideal destination for race-shifting because the tribe has a history of cultural adaptation, tribal exogamy, and relatively open standards of tribal citizenship."[29]

But overall, in contradistinction to Sturm's race shifters, very few individuals on-list identified primarily as Native American or as affiliated with a particular tribe. Those that did identify as Native American, without exception, also descended from European groups and had spent considerable time researching their European ancestry. Most, if not all, of them identified racially as "white," "Caucasian," or "European-American." Some may have identified as "mixed-race," although I do not recall it being a common self-identification. On-list, it was clearly possible to have recent Native American ancestry and still identify as white. This represents a conceptualization of relationships between race categories, that is, between whiteness and "redness," that cannot be explained solely by reference to the binary that exists between black and white in the United States.

Those who sought Native American–ancestry information tended to fall into two categories. First were those individuals who identified as white and who might or might not have Native American ancestry. They hoped a DNA test would provide them an answer, but they identified racially as white and not Native American, whatever the outcome of DNA testing. Second, there were individuals who identified as both white and Native American and who continued to identify as such, whatever the outcome of their DNA testing. Active listers tended to understand ancestry DNA as not providing conclusive evidence for or against Native American ancestry and therefore as not providing a stamp of approval on racial or ethnic identity. I cannot know how less-active listers, or "lurkers," incorporated Native American–ancestry–testing outcomes into their identity reckoning, but active lister responses contradicted what we've seen in some media coverage.[30] Whatever the outcomes of DNA testing, listers' racial-ethnic identifications seemed to stay the same.

This surprised me. When I came to this project, I had hypothesized a researcher's linear progression moving from ancestor questions, to DNA testing, to (for some) affirmative evidence of Native American ancestry, to confirmation of a Native American racial identity. The press has covered identities rescripted after DNA testing, but I formed my hypothesis based more on my lived experience in Indian Country. For years, it has been common to joke about the "Cherokee" or "Indian Princess" phenomenon, in which stories of a long-ago Native American ancestor—confirmed or not—lead individuals to (re)script their identities as Native American, most often Cherokee. Given that DNA tests can neither account for all ancestors nor always differentiate Native American from Asian ancestry, I fully expected that lack of evidence for Native American genetic ancestors would not undermine already held Native identities. Lack of evidence doesn't mean the ancestor wasn't there, just that the particular test in use cannot read the tester's genetic signature. But I did expect that some testers would "race shift" to a newly formed Native American racial identity after testing revealed or confirmed Native American ancestry. To the contrary, they did not. Those who identified primarily as white, Caucasian, or of various European lineages (especially active listers with a good deal of scientific understanding) seemed to emerge from Native American–DNA–testing experiences with white racial identities intact. I found that listers had little trouble reconciling the possibility of Native American ancestry with their whiteness.

In this observation, the listers are both similar to and different from Sturm's shifters. The whiteness of both listers and shifters facilitated their simultaneous claims to Native Americanness. But Sturm's shifters knew their own whiteness as "guilt, loneliness, isolation, and a gnawing sense of racial, spiritual, and cultural emptiness." In their search for fulfillment, they focused on their actual or perceived (and very often "thin") Cherokee ancestral lines as informing their total identities. They sought "transcendence" and conversion from their whiteness.[31] Active listers, by contrast, did not disdain their whiteness. Moreover, they derived deep intellectual and cultural fulfillment through a full and deeply rigorous (social and genetic) scientific exploration of their multiple lineages. I cannot treat their similarities and differences at length here, but a key difference between Sturm's subjects and mine seems to be the communities that have provided them communal experiences on their respective roads to self-actualization and fulfillment. On the one hand, Sturm writes of the race shifters "co-creating a satisfactory spiritual community for themselves and others who have shared their racial journey," of how they "form new Cherokee tribes that provide a context for them to be Cherokee and act Cherokee within a community of like-minded believers. As these new tribes come into being . . . they are a contemporary form of politics that offers both civic engagement and communal spirit in a combination that allows racial shifters to find profound remedies for the ills of the modern, neoliberal age."[32] Listers, on the other hand, seek the benefits of the modern, neoliberal age. They actively build their subjectivities and visibly derive fulfillment as rational and rigorous thinkers within a broader genealogical and scientific community, and as "DNA cousins" with other genetic-lineage-minded folk from across the nation and around the world.

What do these dynamics say about ideas of Native American race on-list and the relationship between Native American ancestry and whiteness? Did listers actually understand Native American identity as do I (being a tribal citizen, former reservation resident, and indigenous studies scholar)? Dare I hope that they understood it as a political designation engendered of tribal-nation sovereignty and governance rights, and therefore not legitimately claimed based solely on a genetic-ancestry test? There were indeed several listers who seemed to understand the nature of tribal and First Nations "citizenship" and who could articulate it as a category other than "race." This may have been a result of their generally good research skills. Those who learned the science may also have been very persistent genealogists and thus had already encountered the enrollment regulations and documentation of tribal governments. But most listers did not seem very aware of the nature of tribal citizenship and did not talk of being Native American in terms of a political designation. Nor did listers tend to talk about Native Americanness in terms of "race." Rather, not surprisingly, they talked about it in terms of "ancestry," "heritage," "affiliation," and "descent," roughly in that order of popularity, or as modifying various genetics terms: "Native American mtDNA," "Native American Y-DNA," "Native American markers,""Native American haplogroups," and the like. But what do such invocations mean? They need to be located within the broader context of how Native American as a category is constituted in the United States, where most listers who claimed Native American ancestry reside.

Native American Heritage as the Legal, Cultural, and Genetic Patrimony of Whites

White Americans make claims to Native American genetic ancestry and identity in ways that mirror the kinds of claims that whites have made to other forms of Native American patrimony—whether land, resources, remains, or cultural artifacts. As legal scholar Cheryl Harris has shown, connections between whiteness and property—both whiteness as a precondition for claiming property and whiteness itself as a valuable form of property—undergird the development of the United States. Furthermore, the property rights accorded to whiteness are protected by the U.S. legal system.[33] One of those rights is control of the legal meanings of group identities. Whites have legally defined who counts as black or Indian. This is an important right, for the racialization and subordination of those black and red "others" has been necessary to solidify the exclusive parameters of whiteness.

In addition to the law, the biological technosciences are becoming increasingly important in the exercise of property claims that sustain our racial formations. Natural resources, including both black bodies and labor and Native American land and resources, constituted important forms of property in earlier centuries that drove the nation-state's civilizing project. In our twenty-first century "knowledge society," the biotechnosciences mine human bodies for raw materials and produce knowledge of them. Both DNA and the knowledge produced of it constitute important forms of property today and drive the current civilizing project. As I have argued elsewhere with Jenny Reardon, Native American DNA, in particular, has emerged as a new natural resource that, like Native American land in the nineteenth century, can be appropriated by the modern subject—the self-identified European, both the scientist and the genealogical researcher—to develop knowledge for the good of the greater society.[34] Through the biotechnosciences, Native American biologies become part of the property inheritance of whites, including the right to use DNA to control the meaning of group identity, or race.

Thus, both law and the biosciences are preoccupied with in-group inheritance. As Yael Ben-zvi explains, "Inheritance connects individuals or generations within particular groups so that biological and material properties are transferred from the deceased to the living members of the same group."[35] Both law and the biosciences then confront the same fundamental question: what constitutes "members of the same group" for the purpose of understanding the transfer of properties? In law, one must determine who is a group member in order to determine who inherits material property. In the biosciences, the order of cause and effect is reversed. Who inherits biological property determines who is a group member.

Paralleling white-supremacist law and presaging twenty-first-century biotechnoscientific claims, nineteenth- and early twentieth-century anthropological theories—Lewis Henry Morgan's theory of cultural evolution being exemplary—facilitated anthropologists' claiming as their inheritance (and the inheritance of the nation) Native American material and cultural patrimony. With American Indian tribes seen as slightly higher on the rungs of the ladder of human evolution than blacks, who were portrayed as falling near the bottom, just above apes, and as largely incapable of further evolution, American Indians were made to represent all of humankind in an early stage of evolution.[36] This enabled a scientific narrative in which whites did not colonize and displace Native peoples but, rather, represented a more evolved form of the same people, "Americans." Native Americans were in turn rescripted as the "vanishing ancestors of their presumably white heirs," who represented the evolutionary pinnacle.[37] Thus, the "institutions, arts, inventions and practical experience" of the Indians formed "part of the human record," and as peoples the Indians possessed "a high and special value reaching far beyond the Indian race itself."[38] The (white) human sciences needed to study the American Indian in order to progress and, as Morgan argued, in order to "recover some portion of the lost history of our own race."[39]

Today, indigenous people, including American Indians, are still rightful objects of study. For Lewis Henry Morgan, American Indian material culture and lifeways were set within "American" history and therefore claimed as part of a (white) American past. American Indian land was then rightfully inherited by U.S. whites.[40] For twenty-first-century researchers, including genealogists, indigenous DNA is part of modern humans' inheritance and a mechanism through which whites can both claim continuity with an aboriginal past and produce knowledge that is ultimately of benefit to all humankind. DNA research and other forms of scientific knowledge production are the twenty-first-century civilizing and development project.

In addition to aiding claims to Native American patrimony, the (potential) absorption of "red" into the white body affected another key shift in U.S. history. As I noted in chapter 1, it helped to consolidate the black/white national race binary, overriding the previous dominant model of race in the United States which had comprised "red, white, and black."[41] Indeed, Native American race, or the race category of "red," was largely neglected in seminal twentieth-century U.S. racial-formation analyses, as scholars commonly focused on the line between white and black.[42] Understanding the consolidation of the black–white binary is critical for understanding histories of race and power in the United States. But then so are the relationships between red and white, and red and black, for that matter. Demarcated less deeply after the late nineteenth century in both anthropology and federal Indian policy, the relationship between red and white has been less visible and, until recently, undertheorized.[43]

Cherokee "Race Shifters" Manifest Whiteness as Property

To clarify how property claims to redness are enabled by whiteness, I want to return to the insights of Circe Sturm and her ethnographic work among individuals, especially from the southeastern United States, who have attempted to shift from a white racial identity to that of an indigenous, specifically Cherokee, identity. She describes a vast and growing movement in which previously identified whites are shifting to Cherokee race identities. Many race shifters have even started or joined "tribes" that are in many cases unrecognized or have gained state but not federal recognition. Race shifters and their tribes are a source of concern for the three recognized Cherokee tribes for reasons having to do with cultural patrimony and the distribution of already scarce government funds to groups they view as fraudulent.[44]

Unlike the listers I encountered, Sturm's subjects seem unwilling to acknowledge or incapable of acknowledging their white racial identity along with their claims to being Cherokee. As I have noted, the listers that did identify as Native American still identified primarily or also as having European ancestry or as "white," "Caucasian," or the like. Listers were keen to explore all of their lineages. Indeed, searching for European ancestries was the primary activity on this list. I also saw listers make no move to forgo a white identity completely in favor of something else. Sturm's race shifters, however, employed the hypodescent norm in their identity reckoning, in which "one drop of Cherokee blood" renders them Cherokee. They were not interested in their European ancestry and associated traditions. That is because Sturm's subjects viewed whiteness as a malady and Indianness as the cure,[45] or whiteness as emptiness and Indianness as the conversion experience that fulfills.[46]

However, what the listers and Sturm's race shifters do seem to share is a "language of choice that they use to describe their racial becoming—without realizing that choice itself is a subtle marker of whiteness." Sturm explains that having a choice about how to racially identify signals "a kind of social power." This social power recalls Harris's description of the privilege of whites to control the legal meanings of group identities not only for themselves but for all others as well. As Sturm explains, race shifters didn't give up their white-skin privilege "but instead relied on it, for whiteness is the mechanism that allowed [them] to reclaim one racial self over another."[47] For both Sturm and Harris, foundational assumptions of power and privilege condition whiteness. Harris explains white control of identity definitions as a mechanism to solidify the exclusive parameters of whiteness. Consistent with this analysis, shifters attempt to control the meaning of Cherokee identity in ways that accord them rights to access Cherokee cultural knowledge. In both Harris's and Sturm's analyses, whiteness invokes rights in identity that facilitate control of patrimony, including land, resources, and cultural practices. Sturm's race shifters often recognize their greater choice in their racial identification. Indeed, they represent their *choice* to be Cherokee as their attainment of a higher "moral standard" than that of those Cherokee who were born into their Cherokee identity.[48]

Another point of comparison is between the biosciences and the race shifters. If anthropology and human-population genetics appropriate indigenous histories to tell the history of all humankind, with modern humans—that is, whites—at the pinnacle, race shifters appropriate

Cherokees as actually white(r) all along. Some shifters believe that Cherokees were not as dark-skinned as other American Indians and that their features were different from those of other Indians. Some even describe Cherokees at the point of contact with Europeans as having blond hair and blue eyes, thus ascribing an "original whiteness" to Cherokee identity.[49] Although the shifter logic of Cherokee racial identity is not synonymous with anthropological logic, both claim American Indian histories and biologies as part of whites' history.

Science as Whiteness

Scholars have shown that race and whiteness change over time.[50] Sturm's race shifters have tired of whiteness. They seek to avoid its perceived cultural emptiness, its lack of spiritual moorings, its "anomic individualism," and, for some in the South, even "shame," and to adopt the rich meanings, traditions, and community they perceive in Indianness. Cherokee identity and practice represent a lost and reclaimed past.[51] Yet the very choice of the shifters in the matter speaks to their partaking of the privileges of whiteness.

Whatever the changes it undergoes, whiteness continues to be associated with modernity. For many, such as genetic-genealogy listers, the modern subject continues to enthrall. They lament the antimodernity of the indigene opposed to genetics. The list and its field of focus, human-population genetics, make clear that today's modern subject is not limited to whiteness in the older way of understanding it, which was tightly tied to certain parts of Europe. Because Europe has long been held out as a beacon of rationality and modernity, and because those traits are now tightly linked to technoscience,[52] individuals deemed in other circumstances to be nonwhite may partake of the whiteness offered up in a scientific subjectivity. "People of color" or "third-world" scientists, for example, can also claim a right to inherit aboriginal cultural and biological patrimony.[53] On-list, as in the field of human genome diversity research, we see the centrality of science in the constitution of a modern, white subject, and we see as well its considerable property claims to genome resources in the name of knowledge production, supposedly for the good of all. This new form of whiteness—constituted through the enactment of a scientific subjectivity—illustrates precisely, I think,

Cheryl Harris's more recent articulation of "post-racial whiteness," in which that race category can no longer be wholly tethered to Europe or to Europeanness.[54]

The next chapter, on the Genographic Project, shows how nineteenth-century anthropological logic continues to ground contemporary genetic science, making it possible to imagine indigenous DNA as a constitutive element of contemporary white bodies. Indigenous DNA and the identities and historical narratives that it enables are thus part of the property that those who can claim a white and modern scientific identity rightfully control.

4

THE GENOGRAPHIC PROJECT

The Business of Research and Representation

In April 2005, the National Geographic Society and IBM, with funding from the Waitt Family Foundation (established by a cofounder of Gateway, Inc.), launched the Genographic Project as a five-year "research partnership"[1] that aims to "trace the migratory history of the human species" and "map how the Earth was populated."[2]

The Genographic Project, a "landmark study of the human journey,"[3] has been frequently compared to the failed Human Genome Diversity Project (the Diversity Project) in both scope and methodological approach.[4] In the early 1990s, a group of scientists proposed a global survey of human genome diversity. They would draw blood from "isolated indigenous populations" that were viewed as highly unadmixed. ("Admixture" refers to the genetic mixing of populations through interbreeding.) This became the Diversity Project. The idea was that such indigenes would provide clear genetic evidence of human evolutionary history.[5] However, sampling was quickly and urgently needed before such "isolates of historic interest" (a term that later rained down trouble from critics) mixed with other populations and evidence of population origins became forever obscured in a murky soup of admixed DNA. The Genographic Project aims to do what the Diversity Project, hampered by controversy, failed to do. And there are more concrete ties between the two projects.

Spencer Wells is the project director and spokesperson for the Genographic Project. He holds a bachelor's degree and a doctorate in biology. He is also a filmmaker, who both masterminded and hosted National Geographic's 2003 documentary *The Journey of Man: The Story of the Human Species,* which makes accessible to nongeneticists a molecular narrative of how humans left Africa sixty thousand years ago to populate the rest of the globe.[6] A charismatic and Nordic-looking spokesperson, Wells has been portrayed on the "Emerging Explorers" page of the National Geographic Web site in his twenty-first-century explorer's uniform, kneeling alongside smaller, nearly naked African hunters who sport bows and arrows.[7] Wells did a postdoctoral fellowship with the Stanford population geneticist Luca Cavalli-Sforza, who founded the Diversity Project. Cavalli-Sforza served as chair of the Genographic Project's advisory board at the project's inception and remains as a board member.[8]

Like the Diversity Project, the Genographic Project consists of teams of scientists from around the world who collect DNA samples, mostly of indigenous peoples, to build a large DNA database—up to one hundred thousand samples, in Genographic's case. Genographic's ultimate goal is synonymous with that of the Diversity Project: to greatly increase the size of the existing database in order to produce a more detailed narrative of human migratory history and the deep historical genetic relationships between different peoples of the world.[9] In addition, both projects have employed the "vanishing indigene" narrative to give a sense of urgency to the drive to collect blood, especially from those who are isolated both genetically and culturally.

Yet Genographic consistently distances itself from the Diversity Project. Clearly conscious of Diversity Project history, Genographic disclaims that it is motivated by the desire to conduct medically relevant research.[10] Diversity Project organizers attempted to ward off the criticism of indigenous peoples with claims about potential health benefits of their research, including the idea that greater understanding of human genetic variation could eventually inform studies of diseases that plague indigenous peoples. Critics were not convinced. They felt indigenous peoples had been duped before into supposedly health-related genetic research that ultimately did not, or that never intended to, address their pressing health issues.[11]

The Diversity Project and the Genographic Project are not alone in generating controversy over genetics research, though the scale of their research attracts much attention. For example, in the 1990s, Arizona State University (ASU) researchers who drew blood samples were accused by some Havasupai tribal members of luring them into a research project focused on diabetes, a condition that occurs very frequently among Native American groups.[12] The research project turned out to be focused on schizophrenia as well, which involved research into historical patterns of consanguinity. Furthermore, samples were transferred to non-ASU researchers who study ancient population migrations. Thus, the indigenous blood samples served three research purposes.[13] The tribe accused researchers of not disclosing the schizophrenia (and consanguinity) research agenda and of aiding population-migrations research. In addition to objecting to the lack of disclosure, the tribal government and tribal members simply didn't want such research, and they eventually sued the university.[14]

In 2002, another indigenous group, the Nuu-chah-nulth First Nation or tribe in British Columbia, Canada, discovered that 883 blood samples taken between 1982 and 1985 by a University of British Columbia geneticist—originally for research on a severely debilitating form of rheumatoid arthritis that occurs at a high rate among them—had been shipped to researchers all over the world. The principal researcher took samples with him (a not uncommon practice) when he assumed positions at the University of Utah and subsequently at the Institute of Biological Anthropology at Oxford University. He continued to do research unrelated to that agreed to on the original consent form, and continued to transfer samples to others. The samples have informed hundreds of academic papers on diverse and controversial topics, including the spread of lymphotropic viruses by intravenous drug use and also research on human migrations.[15] In light of such controversies, it makes sense that the Genographic Project distances itself from population-genetics research that claims health-related benefits for its subjects as it nevertheless moves forward with basically the same research as the Diversity Project.

Genographic also differs from the Diversity Project in important ways that make it more successful than that failed endeavor. Genographic is privately funded. The Diversity Project was funded by the U.S. government and faced public- and tribal-involvement requirements and obstacles that Genographic has largely sidestepped. Genographic is a glossier effort supported by the media power of the National Geographic Society. Photogenic Spencer Wells promotes the project with a fine-tuned ability to weave compelling narratives that support and are supported by genetics and biological anthropology. It is "the stories," he says, that are the reason for doing the research.[16] In the rest of the chapter, I analyze stories that are central to the Genographic Project and to a twenty-first-century discourse of genetic indigeneity that reconfigures older concepts of both indigeneity and biological race. These narratives of population genetics and biological anthropology are not, strictly speaking, untrue, but neither are they the truest or most robust representations of indigeneity and race, either now or in the past. Yet each of the stories highlighted in this chapter gets represented as a fundamental truth about human history and the role of genomics in telling us that history. The culture of genomics is powerful and, in a time of economic crisis, resource-intensive. But although the following stories make claims about the universal value to be found in genome research, they are narratives spun from a particularly narrow standpoint, and thus serve the interests of some in our society and not others. Read uncritically, these narratives are hopeful or inevitable, and they seem multicultural and democratic, but they also imply hierarchical research practices and extractive relations with research subjects, all contextualized within a broader history of colonial violence around the world. These are not simply feel-good stories. They are not possible without histories of violence, and a critical reading from a standpoint that is in part indigenous makes that abundantly clear.

Discover Who You Are (Which Is African), and Genetic Science Will End Racism

It has become a standard claim of human-population genetics that this scientific field can save us from the evils of racism, and Spencer Wells puts it eloquently: "We are all much closely related [*sic*] than we ever expected. Racism is not only socially divisive, but also scientifically incorrect. We are all descendants of people who lived in Africa recently. We are all Africans under the skin."[17] Indeed, this compelling story is at the forefront of Genographic's claims for its relevance. But the story is a bit more complicated than that. The actual work of human-population genetics not only relies on new genetic technologies, it also draws on older notions of biological race and contemporary political understandings of human genetic and cultural diversity. The science does not undermine race and thus racism, but it helps reconfigure both race and indigeneity as genetic categories.

Organizing Genographic's representations is an overarching narrative of global human migration over the past 200,000 years. A half dozen subnarratives operate at multiple levels to support that grand narrative of human movement. But these narratives say much more about the last 500 years of colonial practice and race thinking in the West than they do about the previous 199,500 years of human migratory history. In its principal narrative, the Genographic Project tells participants that by learning more about the migratory routes of their "deep" ancestors, they will "discover . . . who they are and how we are all related."[18] With the popularization of the theory of "Mitochondrial Eve" (mtEve)[19]—a single genetic mother of all living humans—it is common to hear scientists and laypeople claim that we are all really "African."

In a related narrative, "genetic science will end racism," Wells concludes that racism is incompatible with knowledge of genetics. By people's embracing this particular scientific way of knowing the world, racism will be on its way to being extinguished.[20] That is a familiar story: Diversity Project organizers told the same story in arguing for the importance of their project.

In one sense, the first part of the statement, that we are all really African, says nothing. It is nonsensical, given that "Africa" did not exist two hundred thousand years ago when mtEve's DNA inhabited human bodies in the landmass that today we know by that name. In another sense, the statement that we are all really African says a great deal, because Africa is not simply the name given by some humans to a continent. Two long-standing colonial perspectives on Africa are at play. "Africa" cannot be understood outside recent human and colonial history, not even by geneticists or anthropologists. V. Y. Mudimbe, the Zairean-born philosopher, in his now-classic *The Invention of Africa*, has described the colonial narrative embodied in the concept of Africa.[21] This concept embodies difference and primordialism. Africa has long been seen as a place without time or history, a place of irrationality, famine, savagery, violence, and death. It is "the heart of darkness." A statement by Spencer Wells in a *New Zealand Herald* story hints at an "Africa problem" even in ancient human history. Wells explains: "There was a leap forward in intelligence that we see in the fossil and archaeological record about that time, and we think there was strong selection operating. Those that survived [an extreme decline in human population after a catastrophic volcanic eruption] were clever enough to travel and leave Africa." Wells might mean only that humans were clever enough to move elsewhere, but since nonhumans migrate as well, "traveling," as he put it, doesn't require human intelligence. His statement clearly relates cleverness to the drive to leave "Africa."[22] By contrast, Africa has also been seen as "the Rousseauian picture of [a] golden age of perfect liberty, equality and fraternity."[23] Either way, as Mudimbe points out, "Africa" is a loaded concept for Western thinkers. It embodies much more than the notion of a continental landmass out of which came the ancestors of all modern humans.

At first glance, the colonial narrative of African difference, whether as a place of darkness or one of perfect liberty, does not seem to match up with the rhetoric of the Genographic Project and contemporary human genome diversity. Rather, Genographic's rhetoric seems to subvert "otherness" and to celebrate the genetic connectedness of the human species. Indeed, *connected* is a frequently used term in Genographic public relations. But in claiming that "we are all Africans under the skin," Genographic conjures a nineteenth-century racial-science view of connectedness, where Africans preceded the modern white man on the evolutionary chain of humanity. Living Africans, often dressed in "traditional" garb or nearly naked, are used in Genographic imagery and discourse to represent the past, whereas white or "European" subjects are portrayed fully clothed and represent modernity.

Zoologist Peter Dwyer notes that Darwinian evolution "has conditioned an odd understanding: we are what we were, and not what we became."[24] We are all African because our deep genetic ancestors were "African." Genographic does not describe the evolutionary chain of humanity in the oppressive language of race hierarchy. As historians and anthropologists of science have shown, race has not disappeared but has been reconfigured, in part, as "population."[25] Instead of being set in a hierarchy according to a "great chain of being," as races were in the seventeenth through nineteenth centuries, twenty-first-century populations are seen as connected over time and space. We are all one and share the same ancient genetic heritage. With the relatively recent emergence of a multiculturalist ethos in our society that celebrates difference rather than framing it as a hierarchy, the scientific metaphor has changed, in part. As literature and science scholar Priscilla Wald explains, "There is an inherent tension in population genomics between the fluidity of the concept of a population . . . and the methodology and models of the discipline that seem designed to discover its fixity."[26] More recent conceptions of populations as fluid, relational categories sit alongside ideas of discreteness and disparity with surprising ease.

This brings us back to Genographic's second narrative, in which Spencer Wells declares that "racism is scientifically incorrect" because it is incompatible with genetic evidence of connectedness and oneness[27] and that, therefore, genetic science will logically end racism. Wells is wrong on three counts. First, the statement is, in one sense, a non sequitur. Racism does not need to be scientifically "correct" to thrive. Wells's second misstep is related to the first. He is ahistorical when he hopes that scientific knowledge can make the crucial intervention of halting centuries of race oppression. The two sciences chiefly at play here—human-population genetics and (physical) anthropology—have developed in an intimate relationship with dynamic concepts of race and racisms. Histories of racial science demonstrate that fact.[28] Wells either lacks knowledge of that history or he thinks there is discontinuity between the present and the past. Third, Wells privileges one narrative—connectedness—in selling Genographic to the world while simultaneously being informed by ideas of discreteness and disparity that inform human genome diversity research at each stage. Thus, Wells's hope that greater knowledge of our common ancestors will "help people to overcome some of the prejudices they might have" seems naive at best.[29] Greater knowledge hasn't even killed the desire of scientists to seek discreteness and disparity between human groups, and that would seem a simpler task. The role of genetic science in race and racisms is much more complicated than Spencer Wells indicates, as the following narratives demonstrate.

The "Vanishing Indigene"

In giving a sense of urgency to its research program, Genographic reconfigures a centuries-old narrative. The Indian has become the "vanishing indigene," in twenty-first-century parlance. As Spencer Wells claims, "Their numbers are dwindling. Soon they could be gone entirely. We came none too soon."[30] From early in the American colonial era, the Indian was viewed as doomed to vanish under the superior cultural, political, and military force of European and American powers. Some 150 years before Spencer Wells made his similar assertion, Lewis Henry Morgan wrote in *Ancient Society:* "While fossil remains buried in the earth will keep for the future student, the remains of Indian arts, languages and institutions will not. They are perishing daily, and have been perishing for upwards of three centuries. The ethnic life of the Indian tribes is declining under the influence of American civilization, their arts and languages are disappearing, and their institutions are dissolving. . . . These circumstances appeal strongly to Americans to enter this great field and gather its abundant harvest."[31] Since at least the seventeenth century, scientists have joined in such pronouncements. Those who study human physical forms and their biological underpinnings have long warned us that the time is nearly upon us when we will be irrevocably mixed—when our origins will be obscured, when we will no longer be able to glimpse that definitive time of racial purity. Our history—who we are—will be lost to us. The older language of mixing blood has given way to the hastening and inevitable mixing of genes and populations, but the core narrative line remains a powerful ordering device.

Genographic shares common ground with the Diversity Project in seeking to "archive the world's human genetic diversity" before it is too late—that is, before indigenous peoples (those "isolates of historic interest") mix with other populations, making DNA evidence of populational origins muddied forever by (ad)mixed DNA. Blood must be taken before such peoples "lose" their "identity" and "vanish."[32]

As I mentioned in the Introduction to this book, "genetic articulations of indigeneity" rest on assumptions of certain forms of physical and social death. And those assumptions contradict the analyses and work of indigenous peoples who view their presence as peoples, cultures, land stewards, and sometimes governmental entities as still vital. Again, "indigeneity" is articulated precisely in order to defend the *survival* of peoples seen as distinct from settler societies. In indigenous peoples' articulations, that distinctiveness is much more than biological, and it is not a marker of inevitable defeat. Indigenous distinctiveness is a wellspring of cultural, political, and historical strength and the raison d'être of international indigenous movements and law and the expanding fields of Native American and indigenous studies. It is unsurprising, then, that indigenous critics of human genome diversity research resist terms that objectify indigenes as historical-biological curiosities whose raw material must be captured and studied for posterity before they most certainly vanish. DNA sequences—like baskets, bones, pots, and ceremonial objects—become artifacts of soon-to-be-extinct populations. Such a disconnection between genetic and indigenous articulations of indigeneity is not easily bridged.

But genetic scientists can hardly avoid narratives that privilege genetic delineations of identity (and fears of ever-greater admixture) over cultural and political delineations. Although they wade through an ultimately uncategorizable swamp of biogenetic and cultural inputs—and surely most of them know that—their particular research questions and methods compel them to delineate and categorize genetic groupings. Especially since Franz Boas, anthropologists and other scientists have grappled with the notion that racial and, more recently, populational boundaries are confounded by the notion of "culture."[33] Yet the questions that they ask—and, hence, the knowledge that they seek—compel them to know those groups as overly simplified genetic populations. Narrow research questions lead to methods that oversimplify the entanglements of biology and culture, such as when indigenous individuals who are viewed as too highly admixed are eliminated from a sample. The simplification occurs in the following steps:

1. Scientists worry about indigenous peoples vanishing because they view them as storehouses of unique genetic diversity.
2. Since the genetic signatures of "founding populations" are confounded in those who are more highly admixed, those individuals are less useful for research.
3. The admixed indigene becomes not sufficiently indigenous. We see this in common sampling standards, wherein a good research subject should have three or four "indigenous" grandparents, not just one.
4. If admixture is on the rise, the indigene is, by genetic definition, vanishing.

The indigene does not leave genetic inputs out of identity. In a U.S. indigenous context, much of kinship and tribal political citizenship is and always has been biological, but in ways not captured by genetic lineages that are objects of study and analysis. Indigenous assertions of identity and peoplehood turn on complicated intergovernmental policy decisions, laws, collectively held histories, practices, and landscapes, all of which get complexly entangled with biological kinship in a contemporary context. Thus, one might see how the prospect of science showing us "how we are all related," how we got to where we are today, and thus "who we really are" is not necessarily compelling research for indigenes themselves. Such questions do not make sense if peoples already think that they have satisfactory answers to them. The Genographic Project cannot, for example, tell me how I am related to my various Dakota tribal kin, the ultimate set of relations in tribal life. Nor can Genographic tell me how we got here today, although it could tell me that I have the founding "Native American lineage" dubbed "haplogroup A." The question of how we as Dakota got to where we are has already been answered, and the answer does not lie in genetics. I could reference Dakota creation stories that give us values for living, narrate our common history, cohere us as a people with a common moral framework, and tie us to a sacred land base. But another important narrative exists that, for many of us, is arguably more crucial today. We Dakota people got to where we are in the early twenty-first century largely because of what is known in mainstream historiography as the Dakota Conflict of 1862. A full-blown war from the perspective of Dakota historians and community members, the 1862 war recircumscribed present-day Dakota geography, political economy, family relations, governance, and identity.[34] It was the moment when our ancestors' dispossession from our ancestral lands—from the life-giving rivers in what is today southern Minnesota—was crystallized. The Dakotas' pushing back violently against white settlement, and the forced marches, prison camps, and mass execution that ensued, marked a bloody remapping of Dakota life. "Who we really are" is not a question that most, if any, Dakota think can be answered by finding out that they have mtDNA markers that "originated" in Mongolia. All tribal or indigenous peoples have similarly crucial narratives, whether they are creation stories or whether they are those pivotal moments in colonial history that reshaped their lands and thus their land-based identities.

"We Are All Related"

Although admixture is a problem for the research (and for gauging who is sufficiently indigenous for research), admixture is also often framed in a positive light as a "we are all related" story.[35] Given twentieth-century history, this resonates for us in the West. Following eugenics in Nazi Germany, the United States, and Great Britain, and following the civil rights movement and the rise of multiculturalism, the narrative that "we are all related" is important to national cultural histories.

This narrative also has particular resonance for the life sciences that played a controversial role in the race politics of the early twentieth century. After World War II, geneticists decried the race cleansing of Nazi Germany and tried to distance themselves from U.S. complicity in eugenics. Like the notion of the vanishing indigene, the more recent but equally powerful narrative that we are all related is entangled with European and American colonial history, again with particular resonance for geneticists.

This story represents a particular understanding of ancestry, kinship, and self that is culturally and historically contingent. Like DNA-testing companies, human genome diversity research, including the Genographic Project, privileges relatedness along maternal and paternal lines—single genealogical lines—to unnamed genetic ancestors. That makes sense if one's goal is to trace genetic relationships between populations and to approximate dates and geographic direction of human migrations. But matrilines and patrilines are also vested with cultural meanings about origins and (e)migrations, settlement and continental race. When used alone, they entail distinctly nontribal or group ways of reckoning relations and self. They do not illuminate crucial histories, community entanglements, or governance, as those things unfold within the time frame of indigenous memory. Privileging the idea that "we are all related" might be antiracist and all-inclusive in one context, although that is also complicated, because it relies on portraying Africa and Africans as primordial, as the source of all of us. "We are all related" is also inadequate to understanding how indigenous peoples reckon relationships in more complicated ways, both biologically and culturally, at *group* levels.

"We are all related" can also put at risk assertions of indigenous identity and indigenous legal rights. In 1998, Debra Harry and Frank Dukepoo, of the Indigenous Peoples Council on Biocolonialism (IPCB), raised an alarm about this particular issue in relation to the Diversity Project:

> Scientists expect to reconstruct the history of the world's populations by studying genetic variation to determine patterns of human migration. In North America, this research will likely result in the validation of the Bering Strait theory. It's possible these new "scientific findings" concerning our origins can be used to challenge aboriginal rights to territory, resources and self-determination. Indeed, many governments have sanctioned the use of genomic archetypes to help resolve land conflicts and ancestral ownership claims among Tibetans and Chinese, Azeris and Armenians, and Serbs and Croats, as well as those in Poland, Russia, and the Ukraine who claim German citizenship on the grounds that they are ethnic Germans. The secular law in many nations including the United States has long recognized archetypal matching as legitimate techniques for establishing individual identity.[36]

The so-called Kennewick Man controversy shows the potential for human genome diversity research to challenge indigenous identity claims and rights over human remains. When nine-thousand-year-old remains were found near the Columbia River in Washington State in 1996, the first scientist to examine them, James Chatters, assumed they belonged to a Euro-American settler.[37] Carbon-dating analysis soon revealed them to be much older than that, and a group of Native American tribes invoked the Native American Graves Protection and Repatriation Act (NAGPRA), claiming the remains for reburial. The tribes were supported by the Army Corps of Engineers, the agency with jurisdiction over the area where the remains were found.

The involved scientists hoped to disrupt tribal claims to the remains despite their age, by showing that Kennewick Man could not be traced *directly* to contemporary Native Americans. Before repatriation can occur, NAGPRA requires the "cultural affiliation" of remains with living Native American groups. Specifically, the law requires that a "relationship of shared group identity" must be able to be "reasonably traced historically or prehistorically between members of a present-day Indian tribe or Native Hawaiian organization and an identifiable earlier group" via a "preponderance of the evidence—based on geographical, kinship, biological, archeological, anthropological, linguistic, folklore, oral tradition, historical evidence, or other information or expert opinion."[38] With relatively recent remains, burial practices and accompanying material-cultural artifacts can be used to affiliate remains with living

Native American groups whose more recent histories, practices, and kinship ties are documented. But given the age of the Kennewick Man remains, scientists challenged and the court confirmed the material, cultural, and historical evidence as lacking support for "a finding that Kennewick Man is related to *any* particular identifiable group or culture."[39] Prior to the court case, independent forensic analyses had assessed the ancient human as having morphological similarities to several other studied populations, both Native American and Asian. But physical examination was ultimately inconclusive as to whom the remains should be affiliated with in terms of either ancient populations or living populations for which there is data.[40]

Hoping to find a more definitive type of material evidence for Kennewick Man's identity than the forensic data that would disqualify specific tribal claims, the scientists called for genetic analysis of the type used in human genome diversity research.[41] NAGPRA also allows for the documentation of a "lineal-descent" relationship between contemporary Native Americans and the remains or cultural patrimony claimed. Absent cultural evidence, scientists hoped to document this relationship, or its lack, genetically. In the end, however, independent researchers were not able to extract DNA because of mineralization of the bones.[42]

Aside from that technical obstacle, there are three key problems with the reasoning behind using this type of genetic evidence to determine Native American identity in ancient remains. The first is a problem of numbers. The chance that someone living nine thousand years ago has direct genetic descendants living today would be improbable. If we go back only 10 generations, each of us has more than one thousand ancestors. Kennewick Man lived about 450 generations ago. Even if an ancient individual did have direct living descendants, they would likely not be found in the strictly maternal or paternal lineages that can be tested.

The second related problem is with the concept of lineal descent that is key to the implementation of NAGPRA. Lineal descent is a biological concept that is not always compatible with "traditional kinship" concepts or with contemporary ways of determining tribal membership, although, again, Kirsty Gover's analysis of three-hundred-plus tribal constitutions shows that tribes, as they become increasingly genealogical, move toward combining tribal lineal descent with blood rules.[43] Overall, there is no easy and direct connection between lineal descent and tribal belonging. Lineal descent is inadequate to represent the usually more inclusive "traditional kinship systems" of Native American groups of any era for which we have knowledge, and not in line with many contemporary tribal-membership requirements. Yet NAGPRA language seems to view lineal descent and "traditional kinship" as mutually supporting. And NAGPRA applies only to federally recognized tribes, most of which have modern membership requirements that are quite different from "traditional" kinship systems. Yet NAGPRA makes lineal descent central to determining the "cultural affiliation" of remains and cultural patrimony with federally recognized tribes. If reckoned via mitochondrial and Y-chromosome DNA analyses, the concept of lineal descent becomes even narrower and more ill fitting. In the case of the Kennewick Man, had DNA analysis been possible, would lack of proof of a lineal-descent genetic relationship have been used to argue against culturally affiliating Kennewick Man with contemporary Native Americans? Less probable, because of the math and the upheavals of colonial history, genetic evidence might have been found that linked Kennewick Man to living Native Americans—but which ones? In either case, the evidence seems unrelated or, at best, only complementary to other cultural data for the purposes of establishing "cultural affiliation," and not at all relevant for disproving it.

A third related problem with using DNA testing to arbitrate claims to Kennewick Man is that Euro-American colonization troubled the correlations between cultural (as we define them today) and genetic lineages. Of course, definitions of what constitute "admixed" and "non-admixed populations" should be viewed as shifting. Today's ethnic and racial categories, so often used as proxies for genetic populations, were not operative in 1492. It is, therefore, problematic to use them as reference points for defining admixture in relation to ancient human populations. A trip back in a time machine to 1491 would make us redefine admixture (remember that Doerfler's early twentieth-century Anishinaabeg subject would not even have thought in terms of "admixture"); we might perhaps need to refer to something like "belonging" versus "difference"—as between villages, bands, or language groups. What would have been the lines of demarcation and the parameters of inclusion for ancient humans? Certainly not today's continental race categories. In one sense, human movements have always complicated correlations (if people had been able to make them) between cultural and genetic lineages. But to oversimplify and take 1492 as year zero in terms of how we view admixture today in the American continents, processes of settler colonialism have led to shifting racial and cultural identities that are entangled unevenly with the genetic shifts that have also occurred since that time. Individuals from different groups—genetic "populations," "races," "tribes," or "cultures"—have bred with one another. But genetic and social-cultural categories are not pinned perfectly one to another like fabric to a pattern. Even if Kennewick Man had "direct" genetic descendants living today, such shifts make it possible that they would not reside within a Native American group today. Or given that indigenes of the Americas migrated of their own volition before 1492, descendants could be part of native communities far from today's Washington State. The question is, would that be enough evidence to refute specific tribal claims to the remains, and would it bring geographically far-off groups into the mix? With or without genetic evidence, we know that the so-called Kennewick Man walked North America—what is today Washington State—8,500 years prior to European colonization. Verifiable genetic links to the tribes that claim him (living Yakamas and Umatilla, among others), would be very interesting but redundant to that fact. Verifiable genetic links to non-Native Americans anywhere in the world would also be interesting, but they could not be interpreted as conclusive of the lack of a cultural or biological relationship between Kennewick Man and the ancestors of living Native Americans.

Because contradictions between lineal descent and tradition, biology, and culture plague NAGPRA and haunt broader national discussions about Native American identity, we are certain to be confronted again with such conflicts over remains and claims to historical truth. Genetic evidence will rarely give us conclusive evidence of a direct biological relationship of such ancient remains to living, unambiguous Native Americans in the same geographical vicinity. It will mostly further complicate an already complex historical, cultural, and legal field. The worry for indigenous peoples is that genetic definitions of relatedness that inform decisions about "cultural affiliation" will prevail over indigenous definitions and knowledge claims and indigenous human-rights assertions. That was the worry expressed by IPCB back in the early years of the struggle over who governed Kennewick Man. Nearly fifteen years later and despite a 2004 Ninth Circuit Court of Appeals ruling awarding scientists the right to study the remains,[44] the Kennewick Man case is not closed in a broader sense. Scientists, federal agencies, Congress, and Native American tribes continue to assert, legislate, and attempt to regulate who has rights to the remains. The conversation over cultural affiliation and the role of biology in it is no clearer now than it was then. At present, DNA testing is not an option for substantiating or contradicting claims to the remains. However, scientists have noted the possibility that improvements in methods and laboratory techniques will enable successful DNA extraction from remains even as poorly preserved as Kennewick Man.[45] Given that DNA haplogroups organize relatedness at different temporal and spatial levels, "we are all related" can potentially mean something quite different to geneticists and to Native American tribes. Whose definitions will prevail?

Genographic Is a Collaborative Project with Indigenous People

Another key Genographic narrative is its premise that it is a truly collaborative project with indigenous peoples that will help preserve their cultures.[46] This narrative is another way in which the project differentiates itself from efforts, such as the Diversity Project, that loom large in the historical background.

A centerpiece of Genographic's "collaborative" thrust is its Legacy Fund, a grant project to which indigenous groups can apply for funds. Sales of Genographic's Participation Kits (which include painless cheek swabs and "instructions for submitting your DNA samples," plus a keepsake box to store eventual DNA test results in)[47] finance the Legacy Fund, which awards, in most cases, a maximum of $25,000 per grant "for community-driven projects directly preserving or revitalizing indigenous or traditional culture."[48] In 2006, when the Legacy Fund was announced, the IPCB responded to Genographic's call for proposals by alerting its international indigenous network about "the not so altruistic motivations" behind the fund. The council stated that the project provides no "direct benefit to Indigenous peoples and instead raises considerable risks."[49]

The representational risks of the Genographic Project are laid out in detail in this chapter. But any risks—or benefits, for that matter—

resulting directly from the grant making of the Legacy Fund are difficult to assess with the sketchy information provided by Genographic. For most of more than thirty funded projects, only their titles and continent-level locations are available on the Genographic Web site.[50] A handful of projects are located in Africa and the Middle East, one in Europe, eight throughout Asia, two in Australia, one dozen projects across the United States, and a handful in Central and South America. Projects include oral histories, traditional arts and crafts, nearly a dozen language-revitalization and dictionary projects, and other initiatives to document and preserve traditional culture and practices. Several projects incorporate unspecified technologies. There are "community profiles" for only five projects, which provide a bit more detail on the identities of the groups that receive funds, although the terms are still vague. We are told, for example, that the "Traditional People of Gaza" have received funds to revitalize traditional embroidery, and that "the Yagnobi community" (of Tajikistan) receives funds to produce and distribute dictionaries and to hold language classes. The projects are facilitated by NGOs. In line with how Genographic represents "traditional" or "indigenous peoples" in its broader effort, a sense of the diversity of political organization, alliances, perspectives, and projects that exist across indigenous and traditional communities is absent from Genographic's portrayal of its Legacy Fund grant recipients.

Given the goal of indigenous movement—the survival of indigenous peoples as self-determined and self-governing entities—it is unfortunate that Genographic disallows the payment of salaries, "legal actions," travel, conferences, and land acquisition with Legacy Fund grants.[51] Genographic's funding of the peoples it studies is restricted to cultural preservation, a not-unimportant factor for indigenous survival. But indigenous cultural survival cannot be abstracted from land and legal standing, unless, of course, one thinks indigenous cultural survival can be abstracted from living peoples, like artifacts stored on shelves in a museum perhaps.

Genographic's "Indigenous Representatives"

A thirty-minute video featured on Genographic's homepage in 2005 symbolizes Genographic's idea of collaboration, in addition to the Legacy Fund, and Spencer Wells's assertion that integral to the project are the goals of helping to communicate indigenous stories and preserving indigenous cultures.[52] The video, filmed at a project-launch event in April 2005, opens with Spencer Wells alone on-screen. He asks both his live audience and viewers of the video, "What is an indigenous person? What are these indigenous populations we're going to be studying during the course of the project?" Wells is onstage, with a large and impressive projection screen behind him. In talk-show-host style, he invites three men, Genographic's "indigenous representatives," onstage to be interviewed. The indigenous representatives include "a Hadza chieftain from northern Tanzania" (Julias Indaaya Hun/!un//!ume), a Mongolian currently living in San Francisco (Tumur Battur), and a member of the Diné Nation, or Navajo (Phil Bluehouse).

The question of representation looms large. Anyone familiar with indigenous politics will immediately ask, how were the representatives chosen, and by whom? The politics of indigenous representation were a significant barrier to the Human Genome Diversity Project. Diversity Project organizers did revise their initial approach of consulting with nonindigenous experts *on* indigenous peoples, to actually consulting *with* indigenous groups, but they had already done irreversible damage to their project. Their public-relations problem was then magnified as they sought to cultivate real indigenous representation. Determining criteria for representation is no straightforward task. Biomedical ethics on which the Diversity Project drew operate on principles of *individual* consent. Applying those models to groups in order to get *group* consent generated new concerns: How should indigenous groups be chosen in order to get consent? Who decides who is indigenous in order to get participation? Would groups with questionable indigenous status, in effect, be authenticated in order to get consent?[53]

Neither the Genographic Web site nor the video tells us how the featured indigenes were chosen. They are simply presented to us, onstage, sitting before larger-than-life *National Geographic* images of exotic brown faces from around the world. In interviews elsewhere, Wells has noted that he met Tumur (his surname) and Bluehouse through film contacts when he produced *The Journey of Man*, and he knew Indaaya (his surname) from doing fieldwork in Tanzania. This is hardly democratic participation.

On the video, each man (plus an interpreter for Indaaya) ascends the stage and takes his place in one of four empty directors' chairs. An unseen audience applauds. Wells leans forward and asks the three men to tell their stories: "I wanted to give you guys an opportunity to say something about your own people, your own traditions, your own sense of history, your sense of who you are, where you came from, your creation stories, if you will."

The first to speak (through his translator) is the Hadza chieftain, Julius Indaaya Hun/!un//!ume, who describes the daily aspects of his life as a hunter-gatherer. Wells asks him about twenty-first-century pressures on his people. Indaaya responds that another people encroach on their territory and threaten them with loss of land and new diseases with which their traditional medicinal knowledge cannot cope. Indaaya ends his commentary by expressing a desire common to poor people around the world: that their children will be able to access education and will return to help better the community. Wells replies that Indaaya "puts it all in perspective in terms of what [Genographic] is trying to do in terms of raising awareness about these issues," thereby suggesting that Genographic is also a global consciousness-raising project for indigenous struggles.

Further exhibiting the Hadza's cultural distinctiveness, Wells asks Indaaya to translate into the Hadza language "akuna matata," the phrase made famous in Disney's 1994 animated film *The Lion King*. The Hadza speak a clicking language, which Wells emphasizes is "virtually impossible for anyone who hasn't grown up in that society to learn to speak." Impressively, then, explorer-in-residence Wells demonstrates the clicking sounds with perfect (to my untrained ear) elocution. The camera zooms in on Wells in profile as he laughs heartily and appreciatively of Indaaya's indeed beautiful clicking translation. The audience applauds.

Tumur Battur takes center stage. Wearing a t-shirt depicting Genghis Khan, he introduces himself as a Mongolian native and says—with good humor—that his country has a "very good history of Genghis Khan." After all, he continues, Genghis Khan made an empire of half the world. He had many wives and children. Tumur closes his short commentary with the idea that many people may be related to the famous ruler. The camera returns to Wells, who laughs at his indigenous guest's description of Genghis Khan's greatness.

But the next indigene is the star of the show. Diné Phil Bluehouse drives home the main point of the video: Genographic is not a research project that is in opposition to indigenous beliefs and desires; rather, it is in sync both scientifically and spiritually with the indigenous beliefs of the Diné, which is quite an assertion. The Navajo Nation tribal government instituted a genetic-research moratorium within its jurisdictional boundaries in 2001.

Bluehouse's lengthy commentary on the video (mostly in English, although peppered with Diné words) is heartfelt. The following is Bluehouse's account, with minor redundancies eliminated. The account is broken up for ease of analysis, but the passages remain ordered as Bluehouse spoke them:

> My people—the Diné of northeastern Arizona—when we talk about our creation and our subsequent journey— . . . we talk about the creator and how creator thought and how creator speaks and in that planning process and in the delivery of the product—if you want to think about speech as the delivery of the product—we think about how those things integrated with each other and how things formed into what we know now. And we have the narrative about creation, and in that creation there is a divvying up of information and knowledge, and we place that information-knowledge into the sacred colors. One of the sacred colors, we call it the first journey, is the empty color, and then there's the black color, the blue color, the yellow color, and the white color. And each one of those [is] like a database of information that we journey with or [is] contained [Diné word] is what we call them, our scientific terminology for the—maybe the molecules or the cellular structure as to how we are. And then the [Diné word] is that mist form or the cellular structure that [is] independent, and [Diné words] are those things that are within that structure. . . . You might want to call it the genetic strands or something very similar to that.

During the middle of this speech, the camera cuts to Wells, who nods, but his expression is blank. Bluehouse's sentence constructions are difficult to comprehend, but a close review of the video transcription reveals the broad contours of Bluehouse's account. Bluehouse invokes sacred colors, a not-uncommon concept among North American tribes. Sacred colors often refer to directions or seasons, but Bluehouse signifies them uniquely in this narrative. He describes five colors for the Diné: an empty color, black, blue, yellow, and white. He then aligns colors with "journeys," and simultaneously with the concept of molecular "information." He uses computer and molecular metaphors to ascribe meanings to the colors or journeys. In this rendition, sacred colors represent information and knowledge—a "database"—genetic code, molecules, and the cellular structure.

Bluehouse continues:

> So we talk about those things in a very scientific mind, and we describe those events as we're on our journey—the difficulties, the safe times, the good times, the bad times. . . . So we learn and we experience and we progress and we make contact . . . [Diné phrase] is what we say, and we're talking about how migration occurred . . . from the black realm, from the empty to the black, to the blue, to the white, and then the subsequent earth journey, which is the second journey. In those there are many, many oral histories about what happened, who we contacted, who we associated with. We talk about personas, we talk about deities, we talk about relatives all over the place. And when we're talking about those things, we're actually remembering it, because it's already imprinted within us in our DNA and our RNA. So we do chants, we do songs, we do prayers, and all we're doing is reciting those reference points within our existence, and I think that's very exciting to me, because I—I as a human being or as a Navajo or Diné—I am very excited to know who I really am. So when I know who I am, then I feel comfortable with myself, I am at peace with myself. And when you are at peace with yourself, then everything falls in together with peace all around you. And our prayers, we always end with the word [repeats Diné word several times], and we talk about achieving peace, harmony, tranquility, balance, perfection, beauty, and those are the things we do back home.

The creation accounts of indigenous peoples are serious business. They are historically, morally, and spiritually (for lack of better words) crucial to peoplehood. Thus, I make a concerted effort to understand Phil Bluehouse's molecularized description of the creation of the Navajo. But understanding is difficult. His language is characterized by what appear to be non sequiturs. Bluehouse speaks English without a trace of an accent throughout the video. I would not credit his often-convoluted narration to a language barrier. There is probably a strong case to be made about the roughness of any translation of the Diné language or creation stories into English, but that argument does not lend credibility to the idea that genetic and computer-code metaphors generated in the mid to late twentieth century best explain Diné accounts.

I read Bluehouse's narrative with almost no knowledge of Diné creation stories, and I do that deliberately. I want to read the account with the same level of knowledge of Diné tradition that the majority of Genographic's audience could be expected to have. In addition, it is not my place to analyze the veracity of Bluehouse's account as it might be interpreted by living Diné. As a scholar who is Dakota, I share an ethical sensibility with many indigenous people that the traditional knowledge of other peoples is not mine to have. Ethically and practically, I would face challenges in trying to find out how shared or disputed Bluehouse's narrative is among other Navajo, and whether that matters to them. If that is an issue, I think persons enmeshed within their communities should do that work. Rather, my project is to analyze the work that the Bluehouse narrative performs within the national cultural context of the Genographic Project. I trust that there is a fuller, more nuanced account that Bluehouse could provide of how he understands molecular knowledge to be related to the Diné spirit realm and earthly migrations. But this is the account that we have, the one that Bluehouse shared in a public forum. From it, I take the following points of understanding:

1. Diné *creation* is simultaneously a story of *migration* through spirit realms–cum–molecular data sets. Each spirit realm is represented by a sacred color: the empty color, the black realm, the blue realm, the white realm, "and then the subsequent earth journey, which is the second journey." Those sacred colors or spirit realms get resignified as data sets. Each color or realm contains genetic information. Bluehouse explains that Diné "have the narrative about creation, and in that creation there is a divvying up of information and knowledge, and we place that information-knowledge into the sacred colors. . . . And each one of those [is] like a database of information that we journey with." Genetic data and spiritual knowledge go hand in hand in this account, with genetic information partitioned in the act of creation.
2. This creation story is genetically imprinted in Navajo DNA. The chants, songs, and prayers of the people constitute a recitation of those "reference points"—those creation-journey memories that are genetically imprinted. As Bluehouse explains: "There are many, many oral histories about what happened. . . . We talk about personas, we talk

> about deities, we talk about relatives. . . . And when we're talking about those things, we're actually remembering it, because it's already imprinted within us in our DNA and our RNA."

More typically, recitations of stories or histories are understood to enable their transmission to future generations. Bluehouse asserts that the act of recitation is an act of remembering stories that are already encoded genetically. In the Bluehouse account, genetic information enables the recitation rather than the recitation itself being the cause of remembering. As we hear it, genetics and not practice enables Phil Bluehouse to know who he is. This is the account that Genographic broadcast to the world.

Genographic and the Seaconke Wampanoag: Scientific Relevance or Public Relations?

In the Genographic video, Spencer Wells asks, "What is an indigenous person?" The video presents answers in which the desired scientific and cultural attributes meld perfectly. The featured indigenes are sufficiently outwardly indigenous, culturally and phenotypically satisfying. Their Y chromosomes confirm their indigenous identities, their stories, and their indigenous journeys. But what happens when genetic and racial or cultural attributes don't come together seamlessly?

In the summer of 2005, the Seaconke Wampanoag wanted to participate as indigenous research subjects in the Genographic Project. In September and October 2005, newspapers nationally ran a story, which originally appeared in the *Wilkes-Barre (Pa.) Times Leader*, that featured the tribe's participation. Presently located in Massachusetts, the group approached Genographic to have their DNA analyzed, because they wanted more data to supplement limited paper documentation of their history as a tribe and of the genealogy of their members.[54] This article features a photo of Theodore "Tad" Schurr, principal investigator of the North American branch of the Genographic Project, who began taking DNA samples from the Seaconke Wampanoag in August 2005. In the photo, he gazes matter-of-factly into the camera across a painted mask—an indigenous profile—exotic in the foreground.

An article on the Seaconke Wampanoag Web site describes an August 23 visit by Schurr and another scientist, Dr. Sergey Zhadanov, who presented to the Wampanoag an overview of the Genographic

Project and its understanding of human migratory history.[55] In response to the presentation, tribal spiritual leader 3 Bears shared "the story of Turtle Island," presumably an origin story of what we know today as North America, and an audience member suggested the possibility that peoples were also created in North America, with others migrating in later. The meeting is described as a congenial affair.

Sampling began the next morning, on August 24, 2005. Subjects were sampled throughout the day, with the following order being observed: "First those with the closest genalogical [*sic*] connection [to] the root of your respective clan. Second Tribal Leaders, for posterity and historical record. Third any other citizen and selected members with interesting native genealogies."[56] A couple of weeks later, Schurr attended the group's annual powwow, where he took additional blood samples. The article on the tribal Web site ends with the assertion that the tribe was "honored" to be a part of the Genographic study, "to be the first indigenous tribe tested in the North American sector . . . and to lead the way for the indigenous populations of North America."

However, the Seaconke Wampanoag do not enjoy an unambiguous indigenous or "Indian" racial identity. Not unlike the Mashpee Wampanoag, the Massachusetts tribe that James Clifford highlights in his essay "Identity in Mashpee,"[57] the Seaconke Wampanoag appear to be challenged by dominant societal racial images of what constitutes a "real" Indian. Clifford describes the cultural politics at play in courtroom exchanges during a 1977 trial in which the Mashpee Wampanoag Tribal Council sued for the return of about sixteen thousand acres of land, or three-quarters of the town of Mashpee, Massachusetts. Like other eastern tribes, the Mashpee Wampanoag had intermarried extensively with other peoples and did not appear strongly Indian, according to idealized standards. Clifford observes that some members "could pass for black, others for white."[58]

Similarly, judging from photos of tribal council members and other tribal members who appear in the Seaconke Wampanoag online photo album, the group is a virtual rainbow of "admixture." They fall into phenotype categories that range from "Caucasian" to "African" to ambiguous, according to both molecular-anthropological ideas of "population" and according to American racial sensibilities.[59] In their photos, many wear what some might refer to as "traditional regalia," which resemble dance outfits found at powwows all over the country.

The Seaconke Wampanoag are not a federally recognized tribe and thus have not (yet) survived the arduous process that confers federal recognition and its attendant stamp of tribal authenticity and approval. Given the brief historical overviews on the group's home page,[60] the group foregrounds its descent from the Wampanoag who first encountered the Pilgrims in 1621. No genealogical details note any support of this connection, at least on the Web site, which is not surprising. The group laments its lack of genealogical documents and asserts that oral history has been disrupted by colonization. What interests me is that the lack of such evidence is given as a primary reason for engaging in the Genographic Project. The mtDNA and Y-chromosome analyses that the project performs to look for "Native American markers" do not point to specific relations, tribal affiliations, and recent tribal histories. Genographic's DNA analyses cannot tell the Seaconke Wampanoag who they are as Wampanoag, whereas the paper documentation and oral histories, however limited, provide some insight. Genographic's particular research is a bad technical fit for the group's particular needs.

From Genographic's standpoint, the Seaconke Wampanoag would also seem to be a poorly suited technical fit for the project's needs. The *Times Leader* article, like Genographic's project-overview materials, emphasizes Genographic's desire to sample indigenous groups because they are considered less "admixed" than nonindigenous peoples. But certainly the Seaconke Wampanoag, given the obvious admixture of populations in their very recent genealogical history, are not considered good candidates as indigenes by a population-genetics sampling standard. Why, then, when the Seaconke Wampanoag approached Genographic and volunteered to be tested, did Genographic take them up on their offer? And why did the Seaconke Wampanoag go knocking on the door of Genographic instead of pursuing the many other available alternatives for DNA testing? Whatever the particular reasons and despite the technical mismatch, the two parties offer each other a great deal in terms of their respective cultural and public-relations needs.

A 2003 news article described the Seaconke Wampanoag's desires to have land in Rhode Island and Massachusetts returned to them, their willingness to sue, and their plans to seek federal recognition.[61] (The news account also mentions that the tribe would investigate building a casino on land they had acquired, although a more recent news account denies their interest in gaming.) However, James Clifford notes in his analysis of what was at stake in the Mashpee Wampanoag court case, "Although tribal status and Indian identity have long been vague and politically constituted, not just anyone with some native blood or claim to adoption or shared tradition can be an Indian; and not just any Native American group can decide to be a tribe and sue for lost collective lands."[62] Integral to official recognition and land claims is something that is more nebulous but also fundamental, and it is the stuff upon which legal claims may be won: an appropriately authentic Native American image. The cultural image of groups such as the Mashpee or Seaconke Wampanoag *as Indians*—perceptions of them by the broader society as legitimately Native American—is a core issue. Presenting an ungratifying racial image to our highly racialized society presents a challenge to making not only legal claims but also cultural claims, and to personal legitimacy.[63]

By cooperating with Genographic, the Seaconke Wampanoag have received national press as the first group of indigenous people to have their DNA sampled. Being recognized in this way affirms their self-image. Whatever the outcomes of their DNA analyses (which are confidential unless the individuals choose to share them), the Seaconke Wampanoag have been affirmed in the national press *as* indigenous peoples. They accrue indigenous cultural capital, and Genographic is portrayed in the press as "collaborating" with a U.S. tribe.

Like the Genographic video of indigenous representatives, Genographic's relationship with the Seaconke Wampanoag makes for an encouraging multiculturalist, liberal political story. It demonstrates the congenial coming together of indigenous people and geneticists in a mutually beneficial project. The technical incongruence of each party's needs will be lost on most people. Any benefits, however, may come at the risk of subsuming indigenous definitions and meanings to genetic definitions and meanings.

"Native Americans Are Really Mongolians" and "We Are Our Y Chromosomes"

Coming back to the video of indigenous representatives, all three men—the Mongolian, the Hadza chieftain, and the Navajo—have had their Y chromosomes analyzed. It is Spencer Wells's turn now to do the storytelling. One expects to hear a drum roll as he turns to the men and asks them if they would like to hear the results. At this point, the final key narrative comes into play, yet another rendition of "we are what we were." But in this case, it is not that we are all African. Rather, at another temporal level, Native Americans are really Mongolian. This narrative is accomplished through the device of making the Y chromosome and its geographical journey stand in for the man and his journey.

The migratory histories of the three indigenous men—or, rather, that of their Y chromosomes—are revealed for the first time in the video, and are projected onto the gigantic onstage screen. Wells's oration is superior as he takes up the role of the white male scientist-explorer who has just returned from his many travels among the populations and landscapes of the world. He has charted the "journey of man." Fittingly, he shares the stage with only iconographic men. They and their Y-chromosome analyses stand for the journey of indigenes, of those "original" human populations who are now, after thousands of years, on the brink of vanishing into a global sea of admixture.

Wells begins with the Hadza chieftain, and asks for the slide that shows "his journey." A world map in gray and a calming shade of blue appears with the continents labeled. It is entitled "Migration Path, Julius Indaaya, Haplogroup B," as if it was Indaaya himself who made the journey. Presenting the geographical and temporal route of haplogroup B's migration as Indaaya's own migration conveys the idea that DNA markers signify both individual identity and the identity of a people. Wells describes haplogroup B—defined according to the presence of a particular marker, M60—as one of the "most ancient lineages in the Y-chromosome tree," one that points "us back towards Africa as our common origin as a species." The migration path on the slide is shown as a set of arrows moving over the continent of Africa. M60 is "very characteristic of the Hadza people." So, Indaaya is "definitely Hadza, at least on the Y-chromosome side," says Wells, laughing.

Next, Wells reveals Tumur Battur's journey. Tumur is a member of haplogroup C3. Wells tells him and the audience that populational genetic evidence related to C3 reveals that there was "an expansion out of Africa roughly fifty thousand years ago that followed a coastal route along the south coast of Asia." It "reached Australia roughly forty-five to fifty thousand years ago." Wells tells us about subsequent movements of people back into East Asia and Mongolia, all of which can be seen in the distribution of genetic markers around the globe today. He confirms to Tumur that his ancestors indeed came from Mongolia. But there is a surprise. The Mongolian has a special subset of C3 that is "found in high frequency in populations who actually have an oral tradition, where they claim descent from Genghis Khan himself." The high frequency and geographical distribution of the marker, Wells explains to Tumur, "has allowed us to reconstruct what we believe is Genghis Khan's Y chromosome, and you are a direct descendent of Genghis, we think. Congratulations. Just don't kill me." Both men laugh.

There are two problems with Wells's interpretation of Tumur Battur's Y-chromosome analysis. First, he overstretches the evidence when he interprets the data as showing probably that Tumur is a direct descendent of Genghis Khan. We don't have a sample of Genghis Khan's DNA. The evidence indicates only that people who have an oral history of being descended from Genghis Khan share, in high frequency, a male ancestor. The second problem, as with his interpretation of Indaaya's results, is that Wells makes the intercontinental movement over eons of a particular Y-chromosome haplogroup represent Tumur's personal journey. The markers are made to represent the man and, by extension, Mongolians.

Finally, Wells turns to Phil Bluehouse. On the projection screen we see "Migration Path, Philmer Bluehouse, Haplogroup Q." Wells explains that haplogroup Q "delineates the first major expansion into the Americas, within the past twenty thousand years." The camera focuses on Bluehouse, who nods his head and looks into his lap. Wells continues: "These are people who, before they were living in the Americas, were living in Siberia. So you have connections to groups like the Chukchi people and the Tuvinians and Alpines living in southern Siberia. But before that your ancestors were hunting mammoths on the steppes of Central Asia during the last ice age. And before that they were in the Middle East during the second major wave of migration out of Africa around 45,000 years ago. So that is your journey, and you are very typically Native American."

Wells's narrative of the expansion of haplogroup Q into the Americas is indeed fascinating. He weaves scientific evidence into a compelling historical account. His temporal and geographic scope ignites cinematic images in the brain. But again he oversteps the evidence when he authoritatively tells Bluehouse that he is "very typically Native American." What in his account of the migration of haplogroup Q across the continents justifies that conclusion? Haplogroup Q has been dubbed a "Native American marker" for convenience's sake, as shorthand for much more complicated entanglements (never mind the disentanglements) of biology and human "groupness" and movement in the practice of anthropological genetics. But under that label, genetic markers come to stand for Native Americanness. They take on significance beyond their initial designation as markers of ancestry in the Americas found in samples of Native American populations. And remember, for the purposes of historically oriented genetic research, those samples are culled of Native American individuals who are viewed as genetically too admixed. The samples and the markers of interest, then, are not fully representative of the genetic landscape of Native American bodies and populations today. More recent (dis)entanglements of genetics and culture, of genetics and peoplehood, are ignored when Wells takes haplogroup Q not simply as evidence that Bluehouse has genetic ancestors who migrated through today's central Asia and southern Siberia but as evidence that demonstrates Bluehouse's Native American identity.

Not only does the idea that Native Americans are really Mongolian inform Wells's genetic diagnosis of Bluehouse's Native American identity, it is also present in Bluehouse's interpretation of the genetic story. In the question-and-answer session of the video, Bluehouse is brought to tears when an audience member asks him about his reaction to learning about his "ancestors' journey" and the results of his Y-chromosome analysis:

> I guess since . . . childhood I've always had this longing to—to go to a place that I eventually found out was Mongolia. . . . Through my journey—looking at maps when I was in history class in high school . . . it's always been a dream. It's always something that was in me. And finally, I was able to say, "Yeah, it's been confirmed, it's been there genetically—that's what genetics was trying to tell me—that you did come from somewhere." And I think that I did shed tears and it was tears of joy because—[Bluehouse starts crying and takes a moment to collect himself] . . . Making that connection is—I think it's very important.

The camera cuts to Spencer Wells, who nods seriously and swallows, perhaps fighting back his own tears. Bluehouse continues:

> And it goes to the idea of peace and harmony, tranquility and balance, and knowing who we are—so you know in that way [nudging Tumur Battur, who sits next to him, on the leg] this Genghis Khan might be my father! [Laughter and applause onstage and in the audience.] And, of course, my two brothers from Africa [pointing to Julius Indaaya Hun/!un//!ume], you know, somewhere, see, in our creation and journey we talk about first man, first woman. And we're talking about the spiritual realm—things that happen in the mist form [speaks a phrase in Diné]. And we talk about the singularity of being that created all things. And in that creation there was the first man and first woman—the spiritual form, and that spiritual form then started to come together. And I think that describes to us the formation of the cellular structures and the building and the creation that starts to happen after the commandment has been given. So in that sense, I was thinking that somehow we're all just beautifully connected in that way and it's wonderful, I mean, I don't know how else to say it.

Bluehouse's words close the performance.

This video and the event it documents can be seen as a multicultural coming together of genetic and indigenous knowledge, a seamless integration of indigenous and scientific origin stories. Scientists can be quick to discount indigenous origin stories—Wells did it himself in *The Journey of Man* film—yet here he is in good-natured dialogue, laughing with indigenous people, touched by the poignancy of their stories, lending credibility to their oral histories.

But this video—a presentation to the world of Genographic's indigenous "representatives"—does not represent dialogue or democratic participation of indigenes in research. It is a culturally authoritative performance on the part of the Genographic Project. It foregrounds genetic data in the Bluehouse narrative of Diné creation/migration. It notes indigenous struggles and aspirations very briefly, contextualizing them within or losing sight of them against larger-than-life representations of the histories of Y-chromosome migrations. Although Genographic proposes to support indigenous meanings and accounts, it hollows out the meanings of those accounts by attributing to them the "real" meanings, genetic meanings. It confirms indigenous accounts with "real" evidence, genetic evidence. Only when colorfully reinforcing the truth of science, indigenous stories are worthy of being broadcast to the world.

Does Molecularizing Diné Creation Narratives Indigenize the Genographic Project?

Sheila Jasanoff has written that "the production of science and technology becomes entangled with social norms and hierarchies" in "untidy, uneven" ways.[64] Genographic helps give voice to Bluehouse and a molecularized Diné creation account. But it must be noted that there is good reason to believe that Phil Bluehouse's rendition of a Diné creation story represents an evolving understanding of that creation/migration journey. First, Diné creation accounts predate scientific knowledge that has rendered a concise image of our genetic structure in the form of the DNA double helix. Second, creation accounts are far older than the computer-code metaphors that predominate in descriptions of that molecular form. I do not suggest that dynamism in indigenous knowledge is inherently problematic, or that it makes knowledge not indigenous. To the contrary, writing as an indigenous scholar (not as a medicine person or other traditional knowledge keeper), I find evolving narratives encouraging in that they demonstrate that the indigene is not vanishing. Bluehouse, as a Diné, continues to help constitute culture and knowledge.

But does this particular articulation represent a coproduction of indigenous and genetic knowledge prompted too much by the stakes and goals of a corporate genetic-research project? The Bluehouse account bolsters Genographic's claim that it works collaboratively with indigenous people and that its work does not contradict indigenous desires, or even traditional knowledge. But do Diné practices and knowledge shape this constitution of genetic knowledge? Do Diné historical interpretations, moral frameworks, laws, and regulations actually shape the way that Genographic conducts research? These are rhetorical questions, of course. We see no sign that the Genographic Project is "indigenized" in this way. For that to happen, the Navajo nation and Navajo people, not simply Phil Bluehouse as an individual, would have to inform Genographic's work, and in more fundamental ways.

A coproduction analysis is also concerned with "how sociotechnological formations loop back to change the very terms in which we human beings think about ourselves."[65] Regarding Genographic's influence on Diné knowledge, how might formations such as human genome diversity change the way that Diné think about themselves and their history?

We cannot know, from his account alone, the degree to which Bluehouse's involvement in human genome diversity research shapes his narration of Diné creation. That would be the subject of deeper anthropological investigation. Without making Bluehouse—or Genographic, for that matter—an anthropological project, I argue that the individual and the institution are doing mutually constitutive but uneven work that may have implications for indigenous governance as that intersects with genomics. In the Genographic Project, we see a genetic (re)articulation of indigeneity. The deeper contours and nuances of that work are yet to be seen.

An Update on the Genographic–Seaconke Wampanoag Partnership

In early 2010, Genographic Project scientists, along with three Seaconke Wampanoag tribal members, published an article in the *American Journal of Physical Anthropology* that highlighted the potential incongruence between Native American identity and genetic ancestry. The article, "Genetic Heritage and Native Identity of the Seaconke Wampanoag Tribe of Massachusetts," would be more aptly titled "Genetic Heritage versus Native Identity."[66] This paradoxical study, in one sense, represents a superfluous genetic study of the Seaconke Wampanoag tribe's genealogy. Any student of New England tribal history, the tribe's own genealogy, and the Wampanoag people knows that Wampanoag have "mixed" with "European" and "African" populations for a long time. Mitochondrial DNA and Y-chromosome analyses are not necessary to know this history. Indeed, what scientists found confirms existing documentary history: the majority of Seaconke Wampanoag sampled had maternal and paternal lineages traceable to "African" and "West Eurasian" populations.

Nevertheless, depending on how one views the importance of lineal biological descent in constituting a "tribe," it might surprise that the only direct Native American lineage scientists found is traceable probably to a Cherokee ancestor who married a Wampanoag several generations ago.[67] Of course, Native Americans from different tribes traveled the country during the twentieth century, meeting at boarding schools, powwows, and conferences. They migrated in large numbers from rural to urban areas during World War II for employment and, after mid-century, with federal relocation and termination programs. They returned then to reservations with Indian self-determination policies in the 1970s and 1980s. Still, I find it mildly surprising that there is no genetic indication of Wampanoag ancestors in this first genetic analysis. Of course, those ancestors could be on genetic lines not accounted for in the mtDNA and Y-chromosome study.

What I find really surprising—and pleasantly so—in the Genographic–Seaconke Wampanoag article is its tone, which does not conflate Native American identity with Native American genetic lineages. The authors explain that "the high frequency of nonnative haplotypes in this population, along with the paucity of Native American haplotypes, reveals the substantial changes in the genetic composition of the Seaconke Wampanoag Tribe in post-contact American history."[68] This important passage explicitly grounds the subjects as Wampanoag first. Their genetic lineage is not deterministic. This is an interesting turn for the Genographic Project study, given that project discourse up to now has overwhelmingly constituted Native American identity as genetic. In this publication, the project does not trump a preexisting identity with genetic findings that could easily be read as incongruent.

Why this sudden shift in discourse? The article itself—the way it is structured and the names on the byline—provides enough information even without ethnographic inquiry into the scientist-tribe partnership. In a physical-anthropology journal, the authors take the first one-third of the article to recount New England historical literature and the impact of European or white settlement on the numbers and state of the Wampanoag. The authors do the historical accounting in a way that emphasizes Wampanoag survival and not simply their decimation in the face of a brutal colonization. This is a flip of Genographic's usual narrative (and that of the Human Genome Diversity Project before it)—that the indigenes are all vanishing and therefore must be sampled as quickly as possible. The tone of the article is certainly a reflection of the fact that scientists share the byline with tribal community members, a welcome change from the older but still widespread practice of naming tribal subjects—the sample donors—in short acknowledgments at the end of the article. (Even worse, by today's standards, some papers from the early 1990s and before thank agencies such as the Indian Health Service or the Royal Canadian Mounted Police for turning over blood to the scientists with no mention of informed-consent processes in those situations.) The Genographic Project now presents an example of what Louise Fortmann called for,[69] sharing the byline, an important component of collaborative research.

The tribe's collaboration and review of the genetic data presented in this article clearly did not change that data. The genetic findings could be quite damning if presented in a context in which Wampanoag identity was grounded in particular nucleotide sequences. Rather, the publication accounted extensively (for that journal) for colonial history and took care with language. That is not to say that others outside of such a research collaboration would be so nuanced in their approach to interpreting the genetic information presented in the article. The Bureau of Indian Affairs Office of Federal Acknowledgment (OFA), for example, mediates tribal-recognition cases in large part by calling in disciplinarians to pass judgment on the authenticity of Native identity claims. The kind of genetic and *biological* anthropological evidence in the article could easily be brought into the mix in the OFA's deliberations, were a tribe petitioning for federal recognition to present a similar case.

Collaborative publications are not alone sufficient to alter longstanding power relations in knowledge production. Other approaches are necessary in order for indigenous peoples to be real agents of knowledge production. Research questions need to be conceived from indigenous standpoints as well as, or instead of, from nonindigenous-researcher standpoints. In addition, innovative methods need to be developed to account for indigenous moral and epistemological frameworks. Indigenous peoples need to sometimes direct and fund research, and resist it effectively when it is not in their interests. Furthermore, research projects need to be organized and rooted in institutions in which both economic development and institution building accrue to indigenous peoples and their governance and scientific institutions. Still, coauthorship and proper context setting are important first steps.

Conclusion

INDIGENOUS AND GENETIC GOVERNANCE AND KNOWLEDGE

> While perhaps we should not allow particular tribal organizations to yield veto power over scientific studies, it would behoove us as scientists to consult with them as much as possible. Perhaps the genes do not belong to the tribal council—but they certainly do not belong to the scientists.
>
> —Rasmus Nielsen, *Evolutionary Genomics* (blog)

"Native American DNA" fascinated me from the first moment that I heard it uttered. Not having taken a genetics or biological anthropology class, that first utterance struck my ears at a meeting having to do with a grant that my employer had won from the U.S. Department of Energy's program in the ethical, legal, and social implications (ELSI) of genetic research. The three-year grant would enable us to convene tribal representatives together in facilitated sessions to discuss the implications for tribes of mapping of the human genome. I had worked for nearly a decade in the field of tribal environmental programming and policy and was surrounded by people—both Native American and non-Native—who made claims about inherent ("it's in the blood") traditional tribal environmental knowledge and practices. Weary of explanations that drew the focus toward "blood" away from knowledge derived from lived relationships between humans and nonhumans, and humans and place, I was also guarded against a term that seemed to attach Native American practices, history, and identity to strings of DNA molecules. Being able to legitimate one's identity as Native American to the satisfaction of non-Native audiences in the cultural and political theater of U.S. life has become a necessary precondition for asserting rights to tribal self-governance and resources. By engaging in what appears to be biologically essentialist discourse, were we facilitating the tying of those rights to DNA? How would a shift from blood to gene metaphors challenge or infuse the current Native American–rights regime that is rooted in the federal-tribal relationship and federal Indian law and policy?

Given the language at play, it seemed obvious that long-standing race definitions and practices of racialization were hooking into contemporary genome-science knowledges and practices. In the ensuing years, as I studied this problem with both Native American and indigenous studies and science and technology studies frameworks, I came to see this process of influence in a more complex light. Older ideas of race ground new genomic research on human genome diversity. In turn, race concepts and categories that we already know change over time and space were being tweaked and rearticulated by genomic concepts and practices, thus enabling us to conceive of something like "Native American DNA." In short, older race concepts and practices were co-constitutive with genome science.

In addition, when I first came to this work, I did not understand, as I do now, that references to blood might involve not strictly biological aspects but also (or rather) "spiritual" understandings of blood's power that, as Melissa Meyer reminds us, all cultures exhibit over time.[1] To reject such understandings as automatically biologically essentialist is to miss that some blood meanings indeed emerge from a *nonscientific* ethic (I do not mean *unscientific*). Furthermore, using the "science stick," as I've come to think of it, to beat back all blood talk as baseless (it may indeed be essentialist, but perhaps that reflects important cultural ideas) seems ironically to dictate that indigenous peoples live according to biological knowledges that we critical scholars have already claimed they should not be defined and restricted by.

As I've become clearer-thinking about "blood," I have also complicated my understandings about how blood concepts are different from and sometimes overlap gene concepts. Cultural blood meanings will clearly inform Native American adoption of gene language (we are later in coming to gene metaphors than is the mainstream in the United States), and in ways that will not be identical to the use of that language in the mainstream. The transition from blood talk to gene talk and the articulation of concepts such as "genetic memory" in Native American or tribal cultural imaginations is something I have only just begun to approach. This promises to be rich but difficult ethnographic terrain that I may well leave to others, being the reluctant anthropologist that I am.

I initially intended this work to inform debate among tribes in the United States and other indigenous groups about the use of DNA in enrollment and about the risks and opportunities of indigenous involvement (usually as subjects) in academic genetic research. However, this research took unexpected turns during the decade that it was under way. It ended up addressing a broader target than simply tribes. That is the nature of a co-constitution framework: no single phenomenon or discourse can be addressed in isolation. Multiple fields and their norms, as well as multiple understandings of kin, race, and belonging, converge in the object of Native American DNA. I still intend this work be read by tribal eyes. But as it has evolved, I address it no less to nontribal readers. I mostly do not explore the emerging uses of DNA testing in tribal life. Rather, as I explain at length in the Introduction, I have turned my analytical and ethnographic attention to explore the conceptualization and deployment of Native American DNA outside of tribal life in genetic research, in the marketplace of direct-to-consumer genetic-ancestry testing, and in a community of genetic genealogists. I did this for methodological and moral reasons as well as the fact that my sense of the source of the problem with DNA shifted. As I studied the phenomenon, I found myself uneasy with making Native American tribal members my anthropological subjects. This limited my ability to ascertain their no doubt varied and nuanced perceptions of DNA and genetic ancestry. At the same time, it became clear that dominant cultural understandings of kinship and race, more than tribal understandings, give concepts of genetic ancestry their power and salience in our national culture.

Our particular national(ist) history and politics in the United States and the peculiar way in which U.S. indigenes have been racialized in relation to the black–white race binary have made compelling stories of a "Native American" or "Indian" in the family. The U.S. census shows that the number of self-identified Native Americans rose between 1960 and 2000 by 360–650 percent, depending on how they are counted.[2] Tribally enrolled individuals constitute the smaller increase, whereas self-identified Native Americans—especially given that checking multiple race categories was allowed in the 2000 census—constitute the larger growth statistic. Scholars agree that such a rate of growth can be accounted for only by "immigration" and not by natural causes, that is, by births exceeding deaths. Clearly, there is tremendous desire to make the Indian live to serve as an iconic backbone of American identity, whether or not the Indian is a citizen of a nation within a nation. This is why I turned my gaze to the discourses and practices of non–Native American subjects: scientists, genetic-testing companies, and genetic genealogists.

Indigenous Pressure to Reconfigure Genome Research and Policy

For years, indigenous critics have pressed for the reform of physical anthropological and genome-research practices. We now see the fruits of agitation: innovative ethical thinking about how to conduct research in ways that attempt to democratize scientific knowledge production, and an increasing co-constitution of indigenous governance with genome science.[3] In this instance, "democratization" means two things. First, the rights and research priorities of potential subjects are privileged along with the needs and priorities of scientists—or more so when the stakes for subjects are high. These rights include indigenous jurisdiction, or "sovereignty," over research on their lands and knowledges.[4] Second, giving due attention to subject rights and priorities can lead to greater "distributive justice," in which a wider variety of people access a fairer share of the benefits of scientific knowledge production than in the traditional model. Key to that idea is a wider variety of people, such as indigenous communities, (re)defining the benefits to be derived from research by exercising more authority in the process. There are multiple ways to increase that authority, as I will outline briefly. No longer can benefits be restricted to vague promises of possible distant cures to diseases or to knowledge production for the good of all. Indigenous critics see research as not separate from education and training within and without the academy and as not separate from economy—and not only the economy of the scientists' lab, the nonindigenous university, or the state. Individual and tribal institutional-capacity building and economic development should also be viewed as legitimate outcomes of research involving indigenous peoples. Those who want to resist certain research initiatives should also have the tools to do so effectively, and their political and cultural jurisdiction should be respected.

In no small part because of indigenous critic pressures, there are ideas brewing in critical-research communities that tweak scientific questions and methods as well as relations of power between those who study and those who are studied. The most promising strategies for change in genome-research practices—as I and others have argued elsewhere—are not those that are free of state authority but, rather, those that combine the pragmatic advantages of tribal and indigenous regulation with efforts to transform our philosophical and ethical landscapes.[5]

Shortcomings of the Courts and Alternative Mechanisms for Indigenous Governance

The U.S. courts have been a less-than-satisfactory avenue for addressing tribal claims and power inequities related to genomic research or other Native claims to cultural harm based on the mistreatment of Native bodies in research. As American Indian legal scholar Rebecca Tsosie explains, the Havasupai and Kennewick Man cases demonstrate the legal system's inability to recognize tribal claims of both individual and collective spiritual and material harm that can result from research on indigenous bodies and the alienation of biological materials, that is, DNA, from those bodies.[6] Many indigenous groups also believe their rights are coupled with responsibilities to protect the resource.[7] But Euro-American courts value individual autonomy, and they focus on "secular systems of ethics" and property and privacy values that see all resources as capable of being owned, "efficiently" used, exploited, and therefore transferred to ensure more productive uses. They have an easier time recognizing property interests in objects or ideas into which "individual labor and creative effort" have been invested. Indeed, such investment "merits legal protection," whereas DNA is seen as just raw material.[8] The scientist who contributes her intellectual work into the processing and analysis of DNA in the lab, in the U.S. legal paradigm, has the greater property claim.[9]

The cultural myopia conditioning the U.S. legal system prompts Tsosie to call for us to shift the basic theoretical and legal grounds upon which we evaluate indigenous and scientific claims to indigenous genetic resources. She calls for the development of an "intercultural justice" framework that can better promote "intergroup equality" and protect indigenous peoples' human rights. In addition to better accounting for key differences in indigenous approaches to property, privacy, identity, and harm, Tsosie's recommended framework would entail a "restructuring [of] the legal relationships among Native nations and the United States and its non-Indian citizens" to account for historical exploitation of indigenous peoples, including their use as research subjects, that continues to condition power inequities, thus producing harms that continue to affect indigenous communities.[10]

In theory, key differences between indigenous and mainstream values about property, privacy, and identity should enjoy protection, given tribal rights to self-determination. But when challenged, enforcement of tribal sovereignty falls to mediation in state and federal courts, where it is adjudicated according to nontribal cultural conceptions, values, and law.

Drawing on both U.S. tribal and international law, Tsosie recommends mechanisms that can preempt both harms and claims that end up being litigated in non-Native courts. She calls, in large part, for mainstream legal institutional collaboration with tribal court systems and the development of indigenous governance mechanisms that can provide a more effective governance structure for overseeing knowledge production that is not damaging to tribal interests.[11] For example, Tsosie mentions an approach in which researchers sign contracts with indigenous communities that lay out specific conditions under which research will be performed. The Canadian Institutes for Health Research (CIHR) has developed *CIHR Guidelines for Research Involving Aboriginal People*, which advocates such a measure.[12] But tribal jurisdiction is difficult to enforce through mainstream courts when, for example, tribal members live off-reservation or if there is a breach in contract. Thus we see how important are Tsosie's recommendations for cooperation between mainstream and indigenous institutions.

For example, Tsosie calls for the better development of tribal institutional review boards (IRBs) that are positioned to implement tribal values about research and knowledge and that review and set conditions for research on tribal lands. This is a movement that is well under way in Indian Country. But she also calls for mainstream institutions to collaborate better with such tribal institutions to ensure that differing indigenous approaches to property and identity get considered in how we govern research.[13] A good example is a development stemming from the Arizona State University–Havasupai case. After ASU's IRB was shown in the investigative report of that case to have inadequately overseen the research project (the board had information that should have alerted it to the potential for charges of inadequate informed consent), it became much more cooperative with tribal governments on research review. ASU now requires tribal-IRB or tribal-government approval of a protocol that involves any research on tribal lands. Following are several other ideas circulating in critical-research communities that might further Tsosie's call to expand the values and principles according to which we evaluate research protocols and outcomes such that indigenous interests are protected.

Indigenous Control of Biological Specimens

In recent years, indigenous and state agencies in the United States and Canada have put forward promising mechanisms for direct tribal control of biological samples, although they are not without enforcement challenges. The Alaska Area Specimen Bank is Alaska Native-controlled. Located in Anchorage on the Alaska Native Health Campus, the bank is managed by the Alaska Native Tribal Health Consortium (ANTHC). Nine tribal-health organizations make up the ANTHC. During the last fifty years of biomedical research, tribal people served by these health organizations have contributed nearly a half million specimens to the bank.[14] To access specimens, investigators must present research plans in communities whose samples they want to access. After securing community approval for new research, the Alaska Area Indian Health Service IRB must also grant its consent for research on bank specimens.[15] The bank is housed in a Centers for Disease Control and Prevention (CDC) facility as part of a longtime cooperative-research arrangement. Together, tribal health leaders and the CDC developed bank policies and procedures to maximize health benefits to Alaska Natives from any research conducted with samples while protecting Natives' privacy.[16]

A second mechanism for tribal control of biological samples is the DNA-on-loan concept developed by geneticist Laura Arbour and CIHR official Doris Cook. "DNA on loan" means simply that a researcher is considered only a temporary steward of blood and tissues that he or she accepts for research. The community or individual retains ownership and control over the future handling and uses of the samples.[17] The researcher cannot conduct secondary research on the samples without first securing consent for the new research. Anonymized samples, too, are retained as the property of community and individual donors. As long as written consent is obtained stipulating that the samples are "on loan," legal adherence by researchers is required. This model encourages researchers to maintain regular communication and ongoing relationships with communities if they want to make use of samples as new questions and technologies of investigation arise. This approach is opposed to the "helicopter research" that indigenous peoples lament, in which researchers drop in for samples and then leave, never to be heard from again. The DNA-on-loan concept has been the default property arrangement promoted by the *CIHR Guidelines for Health Research Involving Aboriginal People*, which I'll say more about shortly.

Networking Researchers and Changing Disciplines to Promote "Intercultural Justice"

In order for real change to occur that will transform the landscape of genomics research into one of justice for indigenous subjects with their differing concepts of property, identity, privacy, and harm, it will take more than legal institutions making fundamental changes. Scientific institutions will also have to adopt more inclusive cultural frameworks in their governance of genomic and other research.[18]

An interdisciplinary group of scholars, of which I am part,[19] has suggested the need to develop an international research network and clearinghouse that could take some or all of the measures I describe in this section to facilitate the creation and adoption of more intercultural frameworks, as Tsosie suggests. A number of ideas include the international sharing of model codes and contracts,[20] some crafted for use in the United States but revised for potential use among non-U.S. indigenes.[21]

It is also important to highlight the work of critical scientists who are developing new approaches to sampling and genetic-resource governance, such as the DNA-on-loan concept and the tribally controlled biobank. In particular, it is important to call attention to emerging research in which scientists are rethinking their research questions so that they do not reflect only a "Euro-American" view of historical events (including genomic events) and values about which knowledge is important to produce, but address a broader array of standpoints, thus resulting in a broader array of "truths." For example, it will be important for biological anthropologists and other genome scientists, social scientists, and genome-policy experts working in different parts of the world to come together to share, strategize about, and document ways of researching that engage fundamental concepts of race, population, "origins," identity, and the constitution of historical truth more critically and that respond better to indigenous priorities and challenges.

Related to this idea is the need for critical scientists to support educational and advocacy initiatives within national and international scientific associations to suggest changes to professional ethics guidelines and curricula. This can encourage disciplines to be more responsive in their research and teaching to differing concepts of property, identity, and cultural and historical relationships to knowledge.

In addition, innovative educational programs attempt to reconfigure technoscientific and ethical training across the natural and social sciences, humanities, and engineering, thus making a difference in how ethics in research is conceived. One example is Jenny Reardon's brainchild, the Ethics and Justice in Science and Engineering Training Program at the University of California, Santa Cruz (UCSC), in which an interdisciplinary group of scholars and graduate students works under the name of the Science and Justice Working Group on training and research projects.[22] Graduate students—Science and Justice Fellows from science and engineering, humanities, and social-science programs—take seminars, are intensively mentored, and receive funding and research support. They work together, in pairs usually, to "create ethical inquiries" from within and across their own disciplinary practices.

The more usual approach to education and research in the sciences and engineering is to treat ethics and justice concerns as issues to be addressed after research questions are already developed and engineering practices formed. But the UCSC program "trains science and engineering graduate students how to identify and respond to moments within their own research in which good scientific and engineering practices require attentiveness to ethics and justice." It links them to graduate students in the social sciences and humanities and helps students work collaboratively toward just solutions to the problems presented by science and technology in our society. Indeed, the work supported by the UCSC program aims to explode mainstream academic conceptions of ethics that are undergirded by too-narrow cultural philosophies and too-bureaucratic notions of ethics. The interdisciplinary collaboration also benefits social-science and humanities students in that it shows them "how to trace the links between scientific and engineering practices and practices of equity, equality, and power." The program "promises to open up not only novel epistemologies, but new sites and practices for pursuing social justice."[23] The Science and Justice Working Group is also a site where communities, organizations, and industry can bring technoscientific ethical problems they are encountering and have them serve as research foci for the working group.

Changing Federal Agency Guidelines for Research

I have mentioned the *CIHR Guidelines for Health Research Involving Aboriginal People.* The CIHR guidelines, put into effect by 2006, demonstrate the opportunity for federal funding agencies to institute changes that can alter power relations in and avenues for research. In 2012, during final edits to this book, the guidelines were superseded by a newer ethics in funding policy, the second edition of the *Tri-Council Policy Statement: Ethical Conduct for Research Involving Humans* (*TCPS*).[24] The *TCPS* was slated for development prior to the CIHR guidelines, but because it covers not only health research but multiple areas of research and because of bureaucratic delays, this statement took much longer to come into being. In particular, provisions in chapter 9 of the *TCPS* now govern First Nations, Inuit, and Métis health research in Canada and are most applicable to this analysis. With the CIHR guidelines informing them, the *TCPS* ethics are, in practical terms, very similar.[25] As I write, however, the Canadian government is making deep budget cuts to Aboriginal-led health initiatives.[26] What that implies for collaborative health-research efforts with Aboriginal communities in which Aboriginal jurisdiction in the research process is prioritized, is yet unknown. But in a gesture toward optimism, I will continue to refer in the rest of this analysis to the CIHR guidelines and what they do that is ethically innovative.

The CIHR guidelines were described as both "contractual" and "voluntarily assumed by the researcher in return for the funding provided by CIHR."[27] The guidelines are concerned with both political and cultural jurisdiction, as the two are inevitably entangled in health (and genomic) research. There are fifteen articles in all in the document that address key issues that have arisen throughout the chapters of this book. Before the articles are laid out, the guidelines present a grounding assumption that research agreements will be negotiated and signed between researchers and the Aboriginal communities in which they work. Such agreements document mutual understandings of research expectations.

In the first article of the guidelines, researchers are admonished not only to respect Aboriginal "worldviews" as those pertain to notions of collectivism and sacredness of knowledge and specimens, but also to incorporate such language into research agreements. In support of article 1, article 2 states that researchers are expected to understand and respect Aboriginal jurisdiction over research. Indigenous interests in research, then, are understood as not simply cultural interests but also issues of sovereignty and governance.

Research methods are also seen as key to ensuring Aboriginal jurisdiction and cultural relevance and respect. Article 3 states that communities are to be given a choice of a participatory research approach with power sharing in decision making about the content and approach to research in order to ensure cultural appropriateness.

Collective and individual consent are also addressed. Article 4 states that research involving traditional or sacred knowledge, or that targets the community as a collective (for example, research questions that delve into traditional practices or collectively held knowledges or that target the community as a biological population), should seek community-level approval first. Individual informed-consent norms apply after that. Anonymity, privacy, and confidentiality should be respected at both the individual and collective levels. This is an important guideline, as mainstream legal frameworks are geared toward protecting individual privacy and anonymity, often with little concern for the privacy of the collective. Tribes and First Nations have shown themselves to be very concerned with stigmatization of the group.

Akin to the DNA-on-loan idea, which in the guidelines is the default property regime for biological samples, the researcher is expected to understand that the Aboriginal community retains rights to cultural and sacred knowledge, cultural practices, and traditions that have been shared with the researcher during the course of research. The researcher should actively support mechanisms for the protection of such knowledge. Further dissemination of cultural knowledge through recording, video, or written notes—like further dissemination of DNA samples or data—is to occur only with the permission of the community. New research questions put to traditional knowledge data or to biological samples are to be specifically reconsented. In addition, intellectual property rights are to be addressed explicitly in the research agreement precisely because, as Rebecca Tsosie warns, existing laws don't always protect Aboriginal jurisdiction over cultural knowledge. Explicit commercial objectives and/or indirect links to the commercial sector are also to be clearly communicated to Aboriginal communities.

The guidelines also take up the issue of whom research benefits and how. Article 9 states that research should benefit the community and not only the researcher. In addition, benefits—both tangible and intangible—are not to be determined by the researcher but are to be figured by the community.

Related to benefits of research, community-capacity building through education and training should be integral to the research process. In addition, the interactions between Aboriginal communities and educational institutions that ensure power-sharing and accrual of benefits to communities will be enhanced by researchers' learning about Aboriginal protocols in research, by translating to the extent possible all research products into Aboriginal languages, and by prepublication community involvement in data interpretation and review of study conclusions. This type of review is intended not to quash research findings but to ensure appropriate historical and social context of findings and to avoid unnecessary social, cultural, and/or material harm to communities. In the final article, number 15, the Aboriginal community decides how its contributions to the research partnership are to be acknowledged in dissemination of research results. For example, will they negotiate a coauthorship role? Will they be acknowledged in name or anonymously as subjects? This article supports other articles that address multiple concerns with privacy, community benefit, harm, and capacity building.

The take-home points from the guidelines include their recognition of First Nations, Métis, and Inuit peoples in Canada as political entities and not only cultural groups. This is an important insight often missed by non-Native peoples who are more accustomed to thinking in terms of "race" or "ethnicity" rather than indigenous sovereignty. Understanding that indigenous peoples are unique in relation to Euro-American legal and cultural frameworks is key to respecting in practice, and not only in theory, their rights and desires to govern research for the protection of their peoples. There are also indications that collaborative research and power sharing in research governance is not only more just but also more intellectually rigorous. With research assumptions, questions, methods, and analyses being generated from multiple standpoints that work sometimes together and sometimes in tension, greater complexity in knowledge production is a certain outcome.

Finally, the CIHR guidelines are consistent with changes to research approaches advocated by indigenous critics south of the Canadian border for years now. They embody precisely the collaborative move that Tsosie calls for wherein mainstream institutions enact changes that support indigenous political and cultural governance of research in order to better account for differing indigenous concepts of property, privacy, identity, knowledge, harm, and benefits related to research.

Genographic Trouble in Peru—How the CIHR Guidelines Could Help

The potential benefits of ethical guidelines such as those issued by the Canadian Institutes for Health Research are made clear in yet another development in the Genographic Project. In May 2011, Genographic was in the news again for its encounters with indigenous research subjects, but this time the story was not flattering, as it was with the Seaconke Wampanoag. Although the incident figured in only a minor way in the mainstream science press,[28] it lit up the electronic highway connecting my international indigenous networks, both scholarly and activist. A Peruvian NGO, Asociación ANDES, published a ten-page communiqué of the situation, "Genographic Project Hunts the Last Incas."[29] The document offers a comprehensive critique of the Genographic Project and ethical practices surrounding its plans to sample Q'ero people, descendants of Incas who live in a rural area of the Cusco region of Peru.[30] ANDES charges Genographic with being a top-down, primarily extractive research operation with an inadequate informed-consent process that benefits researchers, their institutions, and their economic networks while returning little or no intellectual or economic benefit to their much less powerful indigenous research subjects.

Although we do not have a Genographic response (does Asociación ANDES have facts that Genographic would dispute?), ANDES mounts a dense critique of the project that should be highlighted precisely because the serious charges levied might have been avoided had Genographic adhered to rigorous ethical guidelines like those suggested by the CIHR. ANDES charges Genographic with seven interrelated ethical missteps (which could also be seen as violations of indigenous rights) in relation to research. These missteps are outlined in the following sections.

Inappropriate Consent

ScienceInsider, a publication of the American Association for the Advancement of Science, reported that indigenous leaders in the Cusco region of Peru were charging Genographic scientists with planning to "collect DNA samples without following local regulations and obtaining proper consents," that is, of not getting community input into research plans ahead of time, and of notifying the community only very shortly before planned DNA collection with a patronizing one-page flyer announcing their plans and a PowerPoint presentation *immediately* prior to DNA collection. ANDES writes that Genographic promised in a letter sent to the community "a 'fun' presentation with 'pretty pictures' to induce attendees to offer DNA samples."[31] My colleague Carlos Andrés Barragán, who studies biological-anthropological ethical and discursive practices in Latin America, provided me with a translation of the same letter, which is not on official letterhead, and the author's signature is blocked out. Following is an excerpt: "We are going to use a projector and beautiful images! I am inviting all the Qocha Moqo people (adults, elders, and children!) to participate, because the presentation will be very interesting! Everything is voluntary, there is no obligation, but you will have fun and learn a lot!" Of course, real consent, especially collective consent, takes time. And Genographic's own Web site explains that "informed consent" should be "deliberate, considered, individual and collective." Yet it would seem that Genographic's consent process in Cusco did not allow sufficient time for community input to the research process, and thus precluded "deliberate, considered" collective consent.

Indeed, real informed consent may be impossible without some specialized training in genetics for indigenous peoples (or access to advising by bodies with such knowledge), without which it is difficult to evaluate scientists' claims. Thus, indigenous peoples' capacity building to evaluate the science is key, yet it is apparently absent from the Genographic methodology.

Disrespect of Government Authorities and Lack of Benefits or Harm to Subjects

Some of the informed-consent troubles are linked to ANDES's second charge, that Genographic did not notify local and regional authorities in Peru of their research plans ahead of time, therefore disregarding Cusco's regional-government sovereignty, as well as that of state and indigenous governing bodies. ANDES is unclear in its document about the relations between state, regional, and indigenous governing authorities, and that information seems important to know. States are not always supportive of indigenous self-determination. But there is a hint that those relationships are mutually supportive, at least in this case, in the original letters consulted by *ScienceInsider*.

Carlos Andrés Barragán translated a second letter, written by Benito Machacca Apaza, president of the Hatun Q'eros community (dated April 30, 2011), and addressed to Jorge I. Acurio Tito, president of the Regional Government of Cusco, which reveals the dubiousness of at least some in the community that Genographic's work had any benefit for them, Genographic's third ethical misstep. Indeed, there were worries that it could be harmful. On April 28, the community held a communal assembly, and Machacca Apaza wrote that they "rejected [the Genographic] project because we consider it a great threat to our biological and cultural integrity." He further asked that Acurio Tito, the Cusco region president, enforce the law, request in an expedited fashion more information from Genographic, and "ensure that this people and this project do not harm our culture, our history, our rights to our biological resources, our traditional territories, and our human rights. We want to emphasize that this persons [*sic*] are not welcomed in our communities."

Acurio Tito, in turn, wrote to John Fahey, chairman and CEO of the National Geographic Society, in a letter dated May 2, 2011 (again translated by Barragán), that Genographic had not been granted permission to collect DNA in the community of Qocha Moqo ("Hatun Q'eros"), and that the community "strongly rejects such kind of intervention." Acurio Tito continued: "There has been no consultation process to obtain prior informed consent, as the ILO Convention 169 and the United Nations Declaration on Indigenous Peoples' Rights states it." In addition, "the staff of the Genographic Project has not established any attempt to coordinate these activities with the Regional Government of Cusco," although Cusco regional governmental regulations that seek to "limit bioprospecting" require such consultation. Acurio Tito closes his letter to National Geographic as follows: "In this sense we will not allow a research project that transgresses established regulations; thus, we will apply local, regional, national, and international law to prevent a violation of indigenous peoples' sovereignty and rights."

Who Owns (and Profits from) Biological Samples and Data?

Asociación ANDES also points out Genographic's lack of clarity about future research on and uses of blood samples and sequence data, and disposition of those samples. To what use would samples and data be put? How would reconsent occur if samples were used by non-Genographic researchers for different projects? It is important to be clear on these matters. As ANDES explains, "Highjacking DNA samples collected for one purpose and applying them to another has been a repeated problem with studies on indigenous peoples' DNA."[32] Indeed, it has been common practice for genetic researchers to trade samples and data between labs with no reconsent for different research projects to which sampled groups and individuals might object. After Havasupai, this may be changing. More specifically, in the case of Peru, ANDES and community critics worry that genetic sequences linked to indigenous communities today may in the future be linked to particular medical conditions that can stigmatize indigenous populations *as peoples*. For example, ANDES charges a Genographic scientist from Peru with writing outside of his Genographic work about a particular indigenous group having a "defective gene" that "predisposes them to infectious diseases such as tuberculosis and HIV." Thus, individual anonymity and privacy—the stuff of standard informed-consent models—is inadequate for indigenous groups.

Stewardship of both human and nonhuman biological materials and knowledge is related to charges that some researchers commit "biopiracy" or "biocolonialism." That is, they appropriate indigenous knowledges for the economic and intellectual benefit of nonindigenous researchers and institutions, whereas indigenous groups lose out. ANDES worries that despite Genographic's disclaimer that it has no commercial or medical intent in regard to the samples it collects, individual Genographic scientists "in Peru are active in biomedical and pharmacology research related to indigenous peoples," yet indigenous subjects are not aware of the scientists' previous, potentially stigmatizing publications and ties to commercial enterprise.[33]

Appropriating Cultural and Biological Patrimony to Construct a Genetic Narrative

Biological samples and data are not the only objects of knowledge in which indigenous peoples exercise an interest and which they want to protect. ANDES points out the risks of culturally powerful genetics discourse that often overstates the correlation between genetic ancestry—privileging a very few genetic lineages, in Genographic's case—and individual and group identity. ANDES and the governmental officials in Peru whose correspondence I cite worry that the gene talk of Genographic (and human genome diversity science generally) potentially assaults indigenous identities and historical narratives. ANDES writes that Genographic purports to tell the Q'eros, who proclaim themselves to be "the last Incas," who they really are, even as they struggle to maintain identity and traditions amid the pressures of globalization. Specifically, Genographic will explain "if and how the Q'eros are related to the Incas (as if Inca is defined genetically), related to the Aymara (a neighbouring indigenous linguistic group), or to 'people from the jungle' (i.e., Amazonian peoples)."[34] ANDES's analysis is substantiated in the translation of Genographic's original letter to the community, provided to me by Barragán: "The benefit is that the Q'eros people could know their ancestral roots; in other words, establish if they have a family relation with the Incas, Aymaras, or people from the rainforest. You will be able to learn about your origin centuries back in time."[35] By contrast, Benito Machacca Apaza, president of the Hatun Q'eros community, closed his April 30, 2011, Genographic protest letter to the Cusco regional government as follows: "The Q'ero Nation knows that its history, past, and present, and future is our Inca culture and that we do not need any research called genetics to know who we are. We are Incas, we have always been and we will always be."

The worry is that projects such as Genographic's exist "only to satisfy the curiosity of Western scientists," yet such discursive practices and power will condition indigenous claims to self-governance and rights to land and resources. If human genome diversity scientists "conclude that indigenous peoples are not descended from the original inhabitants of their territories . . . such findings have potential to endanger territorial claims and legal recognitions." Furthermore, "if indigenous territorial claims are so threatened, this opens the door for transnational actors—particularly corporations, which already operate with too few legal and ethical constraints—to move in and begin extracting the natural resource wealth of the region."[36]

Such fears do not seem far-fetched. After all, is it not genetic and other biophysical data, nonindigenous historical narratives, and the moral frameworks of a scientific state that hold sway in dominant courts and institutions? Certainly, indigenous historical narratives, moral frameworks, and data (sometimes biophysical but sometimes immaterial and not knowable by science) do not hold much sway.

It is not only indigenous critics who worry about the power of DNA evidence to constitute history and identity in ways that risk indigenous land and governance rights. UC Berkeley population geneticist Rasmus Nielsen, who studies ancient DNA in the Americas and in Greenland, has blogged that "the predictive power of genetics is hugely exaggerated in the public perception." But "as long as the public [and too many scientists and science reporters, I would add] has faith in the geneticists' ability to predict phenotypes, the implications of disclosing genetic information are enormous."[37]

Sloppy Science

ANDES levies a sixth critique against "molecular biologists," essentially that the science can be sloppy. ANDES explains that scientists' historical claims "sometimes overreach their field of competence and what can ultimately be concluded through science and the historical record. They are influenced by and reliant on assumptions about genetically 'isolated' or 'inbred' populations that discount historical fluidity of cultures and previous intermarriage." Dennis O'Rourke, former president of the American Association of Physical Anthropology, seems to confirm that evidence sometimes gets overextended: "Much of the uncertainty regarding evolutionary inferences of populations relates to the ephemeral nature of populations and the often arbitrary nature of population definitions. This difficulty stems from the fluidity of individual and group identities in time and space. Individuals may change ethnic identities, and hence, group membership, at will, complicating assumptions of demographic continuity over time."[38]

Lack of Full Disclosure

Issues of disclosure are related to problems with consent, disposition of samples, the appropriation of indigenous cultural patrimony, and subpar science. Genographic officially prohibits the commercialization of genetic samples and data by members of its consortium. But ANDES worries that Genographic-affiliated scientists in Peru have broader research programs in which they have pharmaco-genomic interests while also exhibiting in their publications scientific chauvinism about the veracity of indigenous peoples' identities based on their own histories. Therefore, along with calling for clearer lines of accountability for disposition of samples, ANDES calls for Genographic to fully disclose the commercial (and intellectual) interests of its individual scientists, and to more expressly address project efforts to protect indigenous intellectual property in its research program.[39]

ANDES also charges Genographic with focusing to date on "scientifically 'low hanging fruit' (i.e., relatively obvious topics and methods)." Genographic disclaims that it gathers samples for medical research, but given how little it's doing that is scientifically significant under the auspices of this project, ANDES questions this claim. Genographic's main work is to study variations in Y chromosomes and mtDNA, and various medical conditions are linked to mutations in these areas of the human genome. In addition, samples taken by Genographic "contain the full complement of each participant's DNA, and it may be expected that future studies (or current unpublished studies) will expand into analysis of other areas of the genome." ANDES reminds us that Genographic is "notably uninformative about where, for what purposes, and under whose control it will store DNA samples and data for such future uses."[40]

What Difference Would CIHR-Like Guidelines Make?

A brief recounting of CIHR guidelines in relation to Asociación ANDES's critiques reveals that ANDES's ideas are not radical, or antiscience. Rather, they are increasingly recognized as fundamental to ethical research today. CIHR-like guidelines might have helped Genographic and the Q'ero people come to a mutually intelligible understanding of the risks and benefits of human genome diversity research by accounting more fully for political and cultural sovereignties in Peru. Clearly, Genographic's too-vague ethical framework[41] and its various IRB approvals (not all IRBs are created equally) have been insufficiently critical or comprehensive to protect all involved.

The CIHR articles provide guidance for every one of the interrelated issues confronting Genographic in Peru. CIHR admonishes investigators not only to "respect" in theory (as in achieving some multicultural norm to respect someone's right to believe differently, especially when one disagrees) but to actually *understand* Aboriginal worldviews, norms, and jurisdictional authorities, including community jurisdiction. All of this goes beyond simply individual or group *consent* (articles 1, 2, 4, and 11). Consent is part of jurisdiction, but the two concepts are not synonymous.

Related to consent, both individual and community privacy is to be respected (which means stigmatization must be considered) and addressed in a research agreement, as well as indigenous intellectual property rights to cultural knowledges. Research agreements take time to negotiate with communities, and they are a central aspect to the guidelines (articles 5–8). In short, research organizations must study a community's norms and the appropriate governing institutions well ahead of sampling. It seems that Genographic did not take the necessary time to do this background research.

The matter of who has the property interest in biological samples should also be spelled out in research agreements. Absent another agreement, the default framework in the CIHR guidelines was DNA on loan, in which the community owns the biological samples while the researcher is allowed to use them. The transfer of samples and data and their secondary uses require reconsent (articles 12 and 13). ("DNA on loan" is not the official language in the newer *TCPS* guidelines, but the underlying principle of community control and the emphasis on reconsent remain.)[42] Again, we have little to no information on Genographic's plans in this area. Given community concerns about bioprospecting, openness about this aspect of the project plans is crucial.

CIHR also advocates a participatory approach to research and recognizes that real consent is achieved when subjects do more than sign a form and hold out their arms to receive the needle and syringe. Indeed, true collaboration is achieved through the power sharing, continuous two-way sharing of information, and capacity building that come with a participatory approach (articles 3 and 11). To that end, the guidelines specify that education and training of Aboriginal people should be central to research (article 10). There seems to be no doubt that true collaboration is absent from the Genographic scope of work. Related to this is the principle that research should benefit communities and not only researchers. Knowledge for the researcher-defined "good of all" is insufficient. Communities must also define what counts as a benefit (article 9).

Finally, the Aboriginal community should have the opportunity to participate in review of data interpretation and publications to ensure that results are properly contextualized and that sensitive information is not inadvertently revealed to the public. The guidelines are careful to say that this kind of review is not meant to "block the publication of legitimate findings" (article 14). Indeed, Genographic *has* collaboratively evaluated and published research findings with the Seaconke Wampanoag, resulting in the contextualizing of potentially damaging genetic findings. But this kind of approach needs to be spelled out explicitly for the benefit of both researchers and communities.

Toward a Coproduction of Indigenous Governance and Genome Science

On November 3, 2010, I attended, in Washington, D.C., the American Society of Human Genetics (ASHG) Presidential Address by outgoing society president Roderick R. McInnes of McGill University, Canada. The cavernous convention-center exhibit hall was decorated slightly more tastefully than a Las Vegas show or evangelical megachurch: blue and teal cloth was draped from on high to the floor, and a huge ASHG logo and an artsy-looking DNA double helix backlit the plenary-session stage. I sat amid several thousand attentive audience members: genetic scientists, clinicians, and a few ethicists and social scientists, not only from the United States but from around the world. Perhaps more subdued than Vegas showgoers or evangelical Christians (we created a nice low murmur), we sat at the foot of the twenty-first-century altar, the plenary-session stage, where would be paraded before us some of the most interesting new research conducted by both established and young genetic scientists. Our mass was punctuated by several camera operators perched on platforms that enabled them to film and broadcast images of select scientists (race- and gender-diverse, I might add) and their PowerPoint presentations on three gigantic screens at the front of the hall. I had planned to tweet from the session, but because of prepublication privacy concerns, there was no wireless access amid all of the technology. Mobile phones and cameras were forbidden.

Dr. McInnes's talk was entitled "Culture, the Silent Language Geneticists Must Learn to Speak." He began with delightful jokes, at least to my ears and those of the French-speaking individuals next to me, depicting the differences between Americans and Canadians, such as how much better armed we are versus how much better health-insured they are. After cautioning the audience that we should be hesitant to judge genetic studies historically using contemporary ethical standards, he launched into a talk that highlighted genetic-research projects on indigenous groups contested by the researched peoples, including the Nuu-chan-nulth tribe from Vancouver Island, Canada,[43] and the Havasupai.[44] In both cases, consent was obtained and blood drawn for biomedical research but was later used for human-migrations research. In addition, in the case of the Havasupai, stigmatizing schizophrenia research was funded. McInnes also noted the smaller controversy that surrounded the Genographic Project's blood draws in Alaska, in which researchers were asked to return DNA samples until problems about inadequate informed consent were rectified.[45]

McInnes then highlighted indigenous critiques—both activist and scholarly—that have been mounting for years in response to such controversies, critiques that suggest reforms to genetic-research processes and governance. In his thirty-minute address, McInnes provided a synthetic account ranging across literatures outside of his specialty areas, including cultural anthropology, law, genome ethics and policy, and popular science writing. He put together a coherent narrative for non-specialists in genetic-research ethics to explain why the time is now, as he put it, for genetics researchers to "get inside the metaphorical tent of the indigenous populations" they study—to understand their differing approaches to both conceptualizing and governing research. McInnes spoke of something he called "culture" and urged that scientists need to heed the cultural concerns of their indigenous subjects. He highlighted different indigenous cultural beliefs about DNA, for example, noting that some think it is sacred. As late Hopi geneticist Frank Dukepoo put it, "Scientists say it's just DNA. For an Indian, it is not just DNA, it's part of a person, it is sacred, with deep religious significance. It is part of the essence of a person."[46] Genetic findings may also displant the origin narratives of indigenous peoples. McInnes cited me in this instance.[47] To expand beyond the explanation he provided, such narratives give indigenous peoples values for living, narrate our common history, cohere us as peoples with common moral frameworks, and tie us to sacred land bases. Both creation narratives and our own oral and written accounts of colonial events circumscribe our geography, family relations, governance, and identity. Why should genetic knowledge, fascinating as it is, trump these weighty factors?

Beyond cultural concerns, McInnes highlighted the articles of the CIHR guidelines as a different kind of governance structure for health research, including genome research. In addition to being the Alva Chair in Human Genetics at McGill, McInnes was also the scientific director of genetics of the CIHR and was intimately involved in the adoption of those guidelines. He highlighted the need for researchers to respect both Aboriginal political jurisdiction and cultural lifeways, and the need to constitute research benefits from indigenous perspectives as well as researcher perspectives. He cited the Estonian Genome Project, noting that the types of benefits that communities in that country expect to receive from research include better healthcare, better health-care delivery, technology development, economic development, and jobs. "Why," he asked, "should aboriginal populations expect less?"

McInnes's address targeted the "cultural" concerns of indigenous peoples. The title of the address might suggest that geneticists themselves do not have cultural standpoints from which they produce research projects and analyses—a point with which I clearly disagree. I will return to that topic in future conversations with McInnes and other critical scientists. These scientists are struggling to address seriously a wide range of indigenous concerns with genetic research, and I expect the conversations to continue to be fruitful. McInnes's address, in fact, went far beyond "culture" to touch on vitally important governance issues. He suggested reforms that would increase what I see as distributive justice, not only through Euro-American precepts that focus on individual and material benefits but also through indigenous precepts of privacy, collectivity, and sacredness. I heard him suggest reforms that would make genetic science what I would call more democratic and less top-down. Finally, McInnes explained that the lessons learned from research with indigenous peoples and from their efforts to expand their governance of such research are applicable to nonindigenous populations as well. Indigenous peoples bring about innovation in research programs from conception, to implementation, to analysis, to write-up.

"We Are All Related" (Reprise)

Despite cautioning us not to cloak within the palatable moniker of "culture" hard political problems and power-laden historical practices, I return to "culture." I turn to the cultures of both geneticists and indigenous peoples and, perhaps surprisingly, to the words of Native American spoken-word artist John Trudell. In his 2001 album *DNA: Descendant Now Ancestor*, Trudell wields the "we are all related" narrative as he speaks of DNA. However, in his version he invokes Native American ontological notions that expand the number and kinds of beings in relation beyond what traditional genetic-science accounts know to be related. For Trudell, DNA connects not only all humans but all earthly beings (including those not normally thought to contain DNA):

> The DNA of the human. The bone, flesh and blood is literally made up of the metals, minerals and liquids of the Earth. So we are parts of the Earth, we are shapes of the Earth. This is all the things that we are. All the things of the Earth have the same DNA as the human does. Everything of the earth has the same DNA as the human. Everything is made up of the metals, minerals, and liquids of Earth. . . . And being, we have being. That's our essence, that's our spirit. And all the things

> of the Earth have the same DNA as the human [*sic*] have so all the things of the Earth have being and spirit.[48]

Trudell seems to suggest that everything on earth possesses DNA. Although technically this is not true, scientists have seen DNA fragments in meteorites, thus lending credence to the idea that life on earth was seeded from space. It seems, therefore, that Trudell's foundational claim of material connection—whether through DNA or minerals—between all *things* on earth, a claim from which he as a Santee draws a "spiritual" connection, is a robust claim within both Native American and molecular ontologies or worldviews. Trudell's is a deeply connective narrative. I argue that it is more connective than a molecular view that, even if it supports evolution over time and thus continuity between species or beings, still operates too much within a binary framework in which there is life and not-life, humans and nonhumans, culture and nature. In molecular narratives—in gene talk—mutually exclusive and too often hierarchical categories are offered in the same breath as relatedness and connectivity.

Genographic scientist Spencer Wells is avowedly antiracist. Yet he and other academic scientists, when they assert that "we are all related" or "we are all African under the skin," have simultaneously used human-population-genetics methods and narratives to separate and distinguish indigenes and Africans from more modern, less isolated Europeans. In these scientists' research practices, indigenes are clearly approached as less powerful subjects rather than as knowing co-inquirers. Genographic's work is done on material and conceptual ground that leaves intact older race structures in which modern, rational subjects—that is, "Europeans" and scientists—claim rights and privileges to access the resources of those who are seen as less modern, as fundamentally different, and as less evolved genetically.[49] They then claim the right to tell the only "true" story of human history—one in which molecular sequences and their migrations around the world stand in for the lives and movements of human beings, beings who are, of course, constituted of proteins but who could never be defined categorically by those proteins. Indigenous peoples, for example, cohere themselves according to nongenetic criteria when they assert their inherent self-determination as peoples. Peoples also cohere themselves according to narratives that root them in profound and richly historical ways in particular landscapes and landforms, according to key events memorialized in their songs, stories, and ceremonies. Indigenous peoples do not expect scientists to adopt their stories of origin; theirs are not generally proselytizing traditions. But they—we—want our political jurisdictions over our bodies and lands upheld, and we want the power of our stories to shape our lives to be respected and not to be deemed as untruths. The central paradox of the new "global, anti-racist genomics"[50] is that genetic science has advanced historically by violating subjects' rights to self-governance, by appropriating their biological resources (indeed, Genographic has even appropriated indigenous cultural narratives), and by devaluing the truths and powerful values of those it seeks to include and connect.

Genetics discourse can be used to "other," that is, to represent some living human beings as not normative, as the sources of the raw materials of science, as the ancient, remote, less evolved, less enlightened ancestors of more modern living people. Those who are normative are the beneficiaries of science, those for whom scientific texts, films, and truths are produced. Robust genetic data is not absent from the constitution of the narrative that some human groups are ancestral and some are modern. But the narrative of deep genetic and historical connection, responsibility, and less hierarchical social and biological relations among humans and between humans and nonhumans propounded by John Trudell is informed by equally robust material data. It is possible to tell both narratives with some degree of veracity. In the end, we are faced with choices about how to act ethically and socially, about how to treat other human beings and nonhumans. We cannot simply claim that the highest calling is "the" truth, when there are multiple reasonably accurate narratives or ethical and knowledge systems to shape how we might apprehend the world and our obligations to other beings within it.

Beyond "Culture": Power, Politics, and Economy

Why do some narratives and knowledge systems gain more traction than others? It is not simply that scientific experimentation works, and "the" truth wins out in the end. Like other forms of Western knowledge—whether anthropology, philosophy, medicine, history, or the law—genomic practices and data sets cannot be disentangled from histories and politics of resource extraction, or from racisms, colonialisms, and oppressive religious and nationalist doctrines. Science may disrupt certain long-held doctrines, but it has simultaneously been informed and supported by those doctrines and the Western conceptual, political, and material hegemony they justify. Whether we attempt to disown recognition of these politics as an imposition *onto* science or whether we strive to understand how these politics always already condition our scientific practices will make an important difference in whether the genome sciences evolve in more or less democratic ways throughout the twenty-first century.

We indigenous peoples have been forced to confront the sciences and all of the disciplines for the benefit of our communities. We do this to make our and our families' lives more livable, and often because we see such engagement as somehow aiding the survival of our collective peoples. Frankly, we have had little choice but to engage at some level if we are to survive. Science has certainly traded in assimilation, marginalization, and genocide. But it has also been steered toward indigenous goals of self-determination, cultural vitality, and environmental restoration, for example.

It is my hope that more of our leading scientific thinkers will move their own feet, as Rod McInnes suggested they do, into the "metaphorical [I would say 'ontological'] tent" of indigenous peoples or other less powerful subjects whom they target in their research. They don't have to take up citizenship in that new country; they just need to be able to travel and stay there awhile without acting like ugly Americans in a strange and far-off land. They need to resist using their knowledges, techniques, languages, and cultural power as, at best, patronizing, evangelical sermons or, at worst, conceptual sticks to beat down those who are less powerful than them—those who see and think differently. Indeed, I maintain my optimism by seeking out scientists with whom to speak and collaborate who love the revelations and insights their science produces but who also cherish democracy and connectivity. I seek those who are willing to battle within their fields to make space for respectful relations with others who are committed to different but equally moving ways of inhabiting this world. Some of those scientists focus on participatory or collaborative research, and this has been an important step forward.

But it is not the endgame. I also seek out scientists who strive to bring diverse people into their fields, including indigenous scientists, but not just to cast a rainbow across the laboratory. I seek scientific coconspirators who don't think they have the gospel truth but who want to challenge the social and ethical norms of their fields not only in order to make them more inclusive but also to make the science more robust, more strongly objective. They do this in part by recruiting diverse people to their laboratories and then working to make a social and cultural space in which they, too, can flourish. We should look forward to the day that diverse and different-thinking scientists, including indigenous scientists, "colonize" these fields. I use that term as both a biological and a political scientist might use it. Indigenous scientists should move into and settle these fields. But even more, they might extract intellectual and economic resources from research fields—changing them mightily in the process and hopefully for the better (not unlike European colonists thought they were doing over here)—in order also to build contemporary indigenous knowledges, institutions, governance mechanisms, and economies. After all, why should Native Americans and other indigenous peoples expect less? This is the project before us, then: to learn how to grapple explicitly and expertly with the ethics and politics that are always already at the core of the scientific enterprise.

NOTES

Introduction

1. I will refrain from consistently putting scare quotes around "Native American DNA" or referring to it as "so-called Native American DNA," which could be tedious for the reader. But rest assured that I mean the constituted and not simply found nature of that object in every instance.

2. Deborah A. Bolnick et al., "The Science and Business of Genetic Ancestry Testing," *Science*, October 19, 2007, 399–400; Amy Harmon, "DNA Gatherers Hit Snag: Tribes Don't Trust Them," *New York Times*, December 10, 2006; Paul Harris, "The Genes That Build America," *London Observer*, July 15, 2007, 22–27; Brendan Koerner, "Blood Feud," *Wired* 13, no. 9 (September 2005); John Simons, "Out of Africa," *Fortune*, February 19, 2007, 39–43; Takeaway Media Productions, *Motherland: A Genetic Journey*, (London: Takeaway Media Productions, 2003), film; Henry Louis Gates Jr., "The Promise of Freedom," *African American Lives*, episode 2, directed by Leslie Asako Gladsjo, aired February 1, 2006, PBS; and Howard Wolinsky, "Genetic Genealogy Goes Global," *EMBO Reports* 7, no. 11 (2006): 1072–74.

3. Kim TallBear, field notes and personal conversation with legal expert working in this area.

4. Jonathan Marks, "What Is Molecular Anthropology? What Can It Be?" *Evolutionary Anthropology* 11, no. 4 (2002): 131–35.

5. Deborah A. Bolnick, "Individual Ancestry Inference and the Reification of Race as a Biological Phenomenon," in *Revisiting Race in a Genomic Age*, ed. Barbara Koenig, Sandra Soo-Jin Lee, and Sarah Richardson (New Brunswick, N.J.: Rutgers University Press, 2008), 70–88; and Kenneth M. Weiss and Jeffrey C. Long, "Non-Darwinian Estimation: My Ancestors, My Genes' Ancestors," *Genome Research*, no. 19 (2009): 703–10.

6. Sandra Soo-Jin Lee et al., "The Illusive Gold Standard in Genetic Ancestry Testing," *Science*, July 3, 2009, 38–39.

7. Donna Haraway, "Morphing the Order: Flexible Strategies, Feminist Science Studies, and Primate Revisions," in *Primate Encounters*, ed. Shirley Strum and Linda Fedigan (Chicago: University of Chicago Press, 2000), 398–420.

8. Jessica Bardill, "DNA and Tribal Citizenship," *Gene Watch* 23, no. 3 (2010): 8–9.

9. Audra Simpson, "On Ethnographic Refusal: Indigeneity, 'Voice,' and Colonial Citizenship," *Junctures*, no. 9 (December 2007): 67–80.

10. Circe Sturm, *Blood Politics: Race, Culture, and Identity in the Cherokee Nation of Oklahoma* (Berkeley: University of California Press, 2002); and Koerner, "Blood Feud."

11. Charis Thompson, *Making Parents: The Ontological Choreography of Reproductive Technologies* (Cambridge, Mass.: MIT Press, 2005).

12. See Sheila Jasanoff, *States of Knowledge: The Co-production of Science and Social Order* (London: Routledge, 2004); Jenny Reardon, *Race to the Finish: Identity and Governance in an Age of Genomics* (Princeton, N.J.: Princeton University Press, 2005); and Brian Wynne, "Misunderstood Misunderstandings: Social Identities and Public Uptake of Science," in *Misunderstanding Science? The Public Reconstruction of Science and Technology*, ed. Alan Irwin and Brian Wynne (Cambridge: Cambridge University Press, 1996).

13. See Lorraine Daston, *Biographies of Scientific Objects* (Chicago: University of Chicago Press, 2000), for an analysis of the "scientific object."

14. James Clifford, "Indigenous Articulations," *Contemporary Pacific* 13, no. 2 (2001): 468–90; James Clifford, *On the Edges of Anthropology: Interviews* (Chicago: Prickly Paradigm Press, 2003); Stuart Hall, "Gramsci's Relevance for the Study of Race and Ethnicity," *Journal of Communication Inquiry* 10, no. 2 (1986): 5–27; and "On Postmodernism and Articulation: An Interview with Stuart Hall," *Journal of Communication Inquiry* 10, no. 2 (1986): 45–60.

15. Kimberly TallBear, "Native-American-DNA.coms: In Search of Native American Race and Tribe," in Koenig, Lee, and Richardson, *Revisiting Race in a Genomic Age*, 235–52; and Bardill, "DNA and Tribal Citizenship."

16. Vine Deloria Jr., *Custer Died for Your Sins: An Indian Manifesto* (New York: Macmillan, 1969).

17. James Clifford and George E. Marcus, eds., *Writing Culture: The Poetics and Politics of Ethnography* (Berkeley: University of California Press, 1986), 2.

18. Ibid.

19. James Clifford, *Routes: Travel and Translation in the Late Twentieth Century* (Cambridge, Mass.: Harvard University Press, 1997).

20. For example, among many who have treated this topic since the early 1980s, see Emiko Ohnuki-Tierney, "'Native' Anthropologists," *American Ethnologist* 11, no. 3 (1984): 584–86; Kirin Narayan, "How Native Is a 'Native'

Anthropologist?" *American Anthropologist* 95, no. 3 (1993):671–86; Linda Williamson Nelson, "Hands in the Chit'lins: Notes on Native Anthropological Research among African American Women," in *Unrelated Kin: Race and Gender in Women's Personal Narratives*, ed. Gwendolyn Etter-Lewis and Michèle Foster (London: Routledge, 1996), 183–99; Iveta Todorova-Pirgova, "'Native' Anthropologist: On the Bridge or at the Border," *Anthropological Journal on European Cultures* 8, no. 2 (1999): 171–90; Beatrice Medicine, *Learning to Be an Anthropologist and Remaining "Native"* (Urbana: University of Illinois Press, 2001); Lanita Jacobs-Huey, "The Natives Are Gazing and Talking Back: Reviewing the Problematics of Positionality, Voice, and Accountability among 'Native' Anthropologists," *American Anthropologist* 104, no. 3 (2002): 791–804; and Christina Chavez, "Conceptualizing from the Inside: Advantages, Complications, and Demands on Insider Positionality," *Qualitative Report* 13, no. 3 (2008): 474–94.

21. Ohnuki-Tierney, "'Native' Anthropologists," 585.

22. Nelson, "Hands in the Chit'lins."

23. Pakki Chipps, "Family First," *Native Studies Review* 15, no. 2 (2004): 103–5; and Simpson, "On Ethnographic Refusal."

24. Simpson, "On Ethnographic Refusal," 72.

25. Vine Deloria Jr., "Anthropologists and Other Friends," in *Custer Died for Your Sins*, 78–100; and Laura Nader, "Up the Anthropologist—Perspectives Gained from Studying Up," in *Reinventing Anthropology*, ed. Dell Hymes (New York: Vintage, 1972), 289.

26. Paul Robbins, "Research Is Theft: Environmental Inquiry in a Postcolonial World," in *Approaches to Human Geography*, ed. Stuart Aitken and Gill Valentine (London: Sage, 2006), 311–24.

27. Linda Tuhiwai Smith, *Decolonizing Methodologies: Research and Indigenous Peoples* (London: Zed Books, 1999).

28. Marlyn Bennett, "A Review of the Literature on the Benefits and Drawbacks of Participatory Action Research," *First Peoples Child and Family Review* 1, no. 1 (September 2004): 19–32.

29. Rajesh Tandon, "Participatory Research in the Empowerment of People," *Convergence: An International Journal of Adult Education* 14, no. 3 (1981): 20–24; and Bud Hall, "Participatory Research: An Approach for Change," *Convergence: An International Journal of Adult Education* 8, no. 2 (1975): 2431 (cited in Bennett, "Review of the Literature").

30. See, for example, Mark B. Dignan et al., "Health Education to Increase Screening for Cervical Cancer among Lumbee Indian Women in North Carolina," *Health Education Research* 13, no. 4 (December 1998): 545–56; P. A. Fisher and T. J. Ball, "Tribal Participatory Research: Mechanisms of a Collaborative Model," *American Journal of Community Psychology* 32, nos. 3–4 (2003):

207–16; Arizona Biomedical Research Commission, *Community Participatory Research: Enhancing Partnerships with the Native American Community* (Arizona Biomedical Research Commission, 2006); Nina B. Wallerstein and Bonnie Duran, "Using Community-Based Participatory Research to Address Health Disparities," *Health Promotion Practice* 7, no. 3 (July 2006): 312–23; and Sara Goering, Suzanne Holland, and Kelly Fryer-Edwards, "Transforming Genetic Research Practices with Marginalized Communities: A Case for Responsive Justice," *Hastings Center Report* 38, no. 2 (March–April 2008): 43–53.

31. Louise Fortmann, "Gendered Knowledge: Rights and Space in Two Zimbabwe Villages," in *Feminist Political Ecology: Global Issues and Local Experiences*, ed. Dianne Rocheleau, Barbara Thomas-Slayter, and Esther Wangari (London: Routledge, 1996), 220.

32. G. Chidari, F. Chirambaguwa, P. Matsvimbo, A. Mhiripiri, H. Chanakira, J. Chanakira, X. Mutsvangzwa, A. Mvumbe, L. Fortmann, R. Drummond, and N. Nabane, "The Use of Indigenous Trees in Mhondoro District" (occasional paper no. 5, Centre for Applied Social Sciences, University of Zimbabwe, Natural Resource Management, 1992).

33. Fortmann, "Gendered Knowledge," 213.

34. Sergey I. Zhadanov et al., "Genetic Heritage and Native Identity of the Seaconke Wampanoag Tribe of Massachusetts," *American Journal of Physical Anthropology* 142, no. 4 (2010): 579–89.

35. For example, see Bennett, "Review of the Literature"; Bert B. Boyer et al., "Sharing Results from Complex Disease Genetics Studies: A Community Based Participatory Research Approach," *International Journal of Circumpolar Health* 66, no. 1 (2007); Goering, Holland, and Fryer-Edwards, "Transforming Genetic Research Practices"; Gerald V. Mohatt et al., "The Center for Alaska Native Health Research Study: A Community-Based Participatory Research Study of Obesity and Chronic Disease-Related Protective and Risk Factors," *International Journal of Circumpolar Health* 66, no. 1 (2007); and Wallerstein and Duran, "Using Community-Based Participatory Research."

36. Smith, *Decolonizing Methodologies*, 1

37. Native American anthropologist Beatrice Medicine, in *Learning to Be an Anthropologist*, did not position her work as "indigenous methodology" per se, but she, too, noted the importance of taking up topics that are relevant to Native peoples and of doing research that respects and is responsible to family, culture, and tribe.

38. Smith, *Decolonizing Methodologies*, 185, 142.

39. On coupling Maori fundamental assumptions with a university-based methodology, see also Augie Fleras, "'Researching Together Differently': Bridging the Research Paradigm Gap," *Native Studies Review* 15, no. 2 (2004).

40. Smith, *Decolonizing Methodologies*, 188.

41. Ibid., 142–61.

42. Patricia Mariella et al., "Tribally-Driven Participatory Research: State of the Practice and Potential Strategies for the Future," *Journal of Health Disparities Research and Practice* 3, no. 2 (2009): 44.

43. Puneet Kaur Chawla Sahota, "Critical Contexts for Biomedical Research in a Native American Community: Health Care, History, and Community Survival," *American Indian Culture and Research Journal* 36, no. 3 (2012): 1–16; and "Genetic Histories: Native Americans' Accounts of Being at Risk for Diabetes," *Social Studies of Science* 42, no. 6 (2012): 821–42.

44. Eve Tuck, "Suspending Damage: A Letter to Communities," *Harvard Educational Review* 79, no. 3 (2009).

45. Paul Nadasdy, *Hunters and Bureaucrats: Power, Knowledge, and Aboriginal-State Relations in the Southwest Yukon* (Vancouver: UBC Press, 2003).

46. Laurie Anne Whitt, "Indigenous Peoples and the Cultural Politics of Knowledge," in *Issues in Native American Cultural Identity*, ed. Michael K. Green (New York: Lang, 1999).

47. Donna J. Haraway, "Situated Knowledges: The Science Question in Feminism and the Privilege of Partial Perspective," in *Simians, Cyborgs, and Women: The Reinvention of Nature* (New York: Routledge, 1991), 189–91.

48. Ibid., 188.

49. Andrea Nightingale, "A Feminist in the Forest: Situated Knowledges and Mixing Methodism in Natural Resource Management," *ACME: An International E-Journal for Critical Geographies* 2, no. 1 (2003): 77–90; Diane Rocheleau, "Maps, Numbers, Text, and Context: Mixing Methods in Feminist Political Ecology," *Professional Geographer* 47, no. 4 (1995): 458–67; Charis Thompson, *Making Parents: The Ontological Choreography of Reproductive Technologies* (Cambridge, Mass.: MIT Press, 2005); Helen Verran, *Science and an African Logic* (Chicago: University of Chicago Press, 2001); and Jessica K. Weir, *Murray River Country: An Ecological Dialogue with Traditional Owners* (Canberra, Australia: Aboriginal Studies Press, 2009).

50. Sheila Jasanoff, *States of Knowledge: The Co-production of Science and Social Order* (London: Routledge, 2004), 35.

51. Donna J. Haraway, "A Cyborg Manifesto: Science, Technology, and Socialist-Feminism in the Late Twentieth Century," in *Simians, Cyborgs, and Women*, 154.

52. Sandra Harding, *Whose Science? Whose Knowledge? Thinking from Women's Lives* (Ithaca, N.Y.: Cornell University Press, 1991), 109.

53. Sandra Harding, *Sciences from Below: Feminisms, Postcolonialities, and Modernities* (Durham, N.C.: Duke University Press, 2008), 3.

54. Ibid.

55. Harding, *Whose Science? Whose Knowledge?* 124.

56. Roderick R. McInnes, "Culture, the Silent Language Geneticists Must Learn to Speak," *American Journal of Human Genetics* 88 (2011).

57. Rebecca Tsosie, "Cultural Challenges to Biotechnology: Native American Genetic Resources and the Concept of Cultural Harm," *Journal of Law, Medicine, and Ethics* 35, no. 3 (2007).

1. Racial Science, Blood, and DNA

1. Michel Foucault, "Nietzsche, Genealogy, History," in *The Foucault Reader*, ed. Paul Rabinow (New York: Pantheon, 1984), 76–100.

2. Ibid., 81.

3. See, for example, Elazar Barkan, *The Retreat of Scientific Racism* (Cambridge: Cambridge University Press, 1992); Robert E. Bieder, *Science Encounters the Indian, 1820–1880: The Early Years of American Ethnology* (Norman: University of Oklahoma Press, 1986); Jonathan Marks, *Human Biodiversity: Genes, Race, and History* (New York: de Gruyter, 1995); and George Stocking, *Race, Culture, and Evolution: Essays in the History of Anthropology* (New York: Free Press, 1968).

4. I owe thanks to Circe Sturm and her *Blood Politics: Race, Culture, and Identity in the Cherokee Nation of Oklahoma* (Berkeley: University of California Press, 2002) for this term, which I borrow and use in this chapter.

5. Barkan, *Retreat of Scientific Racism;* Nancy Stepan, *The Idea of Race in Science: Great Britain, 1800–1950* (Hamden, Conn.: Archon Books, 1982); and Stocking, *Race, Culture, and Evolution.*

6. David Theo Goldberg, *Racist Culture: Philosophy and the Politics of Meaning* (Oxford: Blackwell, 1993), 49–51.

7. Stocking, *Race, Culture, and Evolution*, 44.

8. Ibid., 30.

9. See, for example, Peter Wade, *Race, Nature, Culture: An Anthropological Perspective* (London: Pluto Press, 2002); and Stepan, *Idea of Race in Science*, 9.

10. Stephen Jay Gould, "American Polygeny and Craniometry before Darwin," in *The "Racial" Economy of Science: Toward a Democratic Future*, ed. Sandra Harding (Bloomington: Indiana University Press, 1993), 85, 89.

11. More recently, the phrase "biogeographical ancestry" is not being widely used in the field. Rather, scientists frequently use simply "ancestry," which carries the same, in part racial, connotations.

12. Marks, *Human Biodiversity*, 120–21.

13. Stepan, *Idea of Race in Science*, 50–51, 84–87; and Stocking, *Race, Culture, and Evolution*, 46.

14. Stocking, *Race, Culture, and Evolution*, 231.

15. Ibid., 199–202.

16. Ibid., 214–31.

17. Barkan, *Retreat of Scientific Racism*.

18. *RACE: Are We So Different?* American Anthropological Association, accessed June 10, 2012, http://understandingrace.org/home.html.

19. See, for example, Mark Shriver et al., "Skin Pigmentation, Biogeographical Ancestry, and Admixture Mapping," *Human Genetics* 112 (2003).

20. Reardon, *Race to the Finish*, 23.

21. Stocking, *Race, Culture, and Evolution*, 167.

22. Allyson D. Polsky, "Blood, Race, and National Identity: Scientific and Popular Discourses," *Journal of Medical Humanities* 23, nos. 3–4 (2002): 178.

23. Stepan, *Idea of Race in Science*, 144.

24. See Evelyn Fox Keller, *The Century of the Gene* (Cambridge, Mass.: Harvard University Press, 2002), 1–3, for a discussion on the speculative meanings of "gene" in early genetics.

25. See Stepan, *Idea of Race in Science*, 172–73.

26. Marks, *Human Biodiversity*, 59–60.

27. Reardon, *Race to the Finish*, 79 and 40.

28. Marks, "What Is Molecular Anthropology?" 131–35.

29. Jonathan Marks, *What It Means to Be 98% Chimpanzee: Apes, People, and Their Genes* (Berkeley: University of California Press, 2001), 41–42.

30. Evelyn Fox Keller, *Refiguring Life: Metaphors of Twentieth-Century Biology* (New York: Columbia University Press, 1995), 93.

31. Ibid., 18.

32. Francisco J. Iborra, Hiroshi Kimura, and Peter R. Cook, "The Functional Organization of Mitochondrial Genomes in Human Cells," *BMC Biology* 2, no. 9 (2004).

33. Elgar Greg and Tanya Vavouri, "Tuning In to the Signals: Noncoding Sequence Conservation in Vertebrate Genomes," *Trends in Genetics* 24, no. 7 (2008).

34. There are pseudoautosomal sections of the Y chromosome that do recombine with the X chromosome. However, Y-ancestry tests focus on markers in the male-specific region.

35. Ripan S. Malhi et al., "The Structure of Diversity within New World Mitochondrial DNA Haplogroups: Implications for the Prehistory of North America," *American Journal of Human Genetics* 70, no. 4 (2002).

36. Stephen L. Zegura et al., "High-Resolution SNPs and Microsatellite Haplotypes Point to a Single, Recent Entry of Native American Y Chromosomes into the Americas," *Molecular Biology and Evolution* 21, no. 1 (2004).

37. Dennis H. O'Rourke and Jennifer A. Raff, "The Human Genetic History of the Americas: The Final Frontier," *Current Biology* 20, no. 4 (2010).

38. Zegura et al., "High-Resolution SNPs."

39. See Megan Smolenyak Smolenyak and Ann Turner, *Trace Your Roots with DNA: Using Genetic Tests to Explore Your Family Tree* (Emmaus, Pa.: Rodale, 2004), 93–96, for a clear description of autosomal markers in genetic-ancestry testing.

40. L. Luca Cavalli-Sforza, Paolo Menozzi, and Alberto Piazza, *The History and Geography of Human Genes* (Princeton, N.J.: Princeton University Press, 1994), 3; and Marks, *Human Biodiversity.*

41. Theodore W. Allen, *The Invention of the White Race*, vol. 2, *The Origin of Racial Oppression in Anglo-America* (London: Verso, 1997); and David Theo Goldberg, *The Racial State* (Malden, Mass.: Blackwell, 2002), 115–16.

42. Yael Ben-zvi, "Where Did Red Go? Lewis Henry Morgan's Evolutionary Inheritance and U.S. Racial Imagination," *CR: The New Centennial Review* 7, no. 2 (2007).

43. Robert Berkhofer Jr., *The White Man's Indian: Images of the American Indian from Columbus to the Present* (New York: Random House, 1978).

44. Ibid.; Brian W. Dippie, *The Vanishing American: White Attitudes and U.S. Indian Policy* (Lawrence: University of Kansas Press, 1991); and Ben-zvi, "Where Did Red Go?"

45. Ben-zvi, "Where Did Red Go?" 202.

46. Dippie, *Vanishing American.*

47. Ben-zvi, "Where Did Red Go?" 212–13.

48. Ibid., 203; Lewis Henry Morgan, *Ancient Society; or, Researches in the Lines of Human Progress from Savagery, through Barbarism to Civilization* (1877; repr., Chicago: Kerr, 1909); and Lewis Henry Morgan, *Houses and House-Life of the American Aborigines* (1881; repr., Chicago: University of Chicago Press, 1965).

49. See also see Morgan, *Ancient Society;* Morgan, *Houses and House-Life;* Gould, "American Polygeny and Craniometry"; and Stocking, *Race, Culture, and Evolution.*

50. Ben-zvi, "Where Did Red Go?" 206–7.

51. Berkhofer, *White Man's Indian;* Vine Deloria Jr. and Clifford M. Lytle, *The Nations Within: The Past and Future of American Indian Sovereignty* (Austin: University of Texas Press, 1984).

52. Piero Camporesi, *Juice of Life: The Symbolic and Magic Significance of Blood* (New York: Continuum, 1995), 27.

53. Ibid., 32. See also Dorothy Nelkin, "Cultural Perspectives on Blood," in *Blood Feuds: AIDS, Blood, and the Politics of Medical Disaster,* ed. Eric Feldman and Ronald Bayer (New York: Oxford University Press, 1999).

54. Jean Dennison, "Constituting an Osage Nation: Histories, Citizenships, and Sovereignties" (PhD diss., University of Florida, 2008), 90–91.

55. Melissa Meyer, *Thicker Than Water: The Origins of Blood as Symbol and Ritual* (New York: Routledge, 2005), 208.

56. Melissa L. Meyer, "American Indian Blood Quantum Requirements: Blood Is Thicker Than Family," in *Over the Edge: Remapping the American West*, ed. Valerie J. Matsumoto and Blake Allmendinger (Berkeley: University of California Press, 1998); and Nelkin, "Cultural Perspectives on Blood."

57. On state uses of race, see Allen, *Origin of Racial Oppression*, 2. See also Goldberg, *Racial State*. For the way scientists and scientific institutions used race to order human bodies and trade in body parts for scientific study, see Harding, *Sciences from Below;* Bieder, *Science Encounters the Indian;* Marks, *Human Biodiversity;* and Deloria, *Custer Died for Your Sins*.

58. Marks, *Human Biodiversity*, 126–33.

59. L. Hirschfeld and H. Hirschfeld, "Serological Differences between the Blood of Different Races," *Lancet*, October 18, 1919, 675–79, in ibid., 127.

60. Marks, *Human Biodiversity*, 127.

61. David Schneider, *American Kinship: A Cultural Account* (1968; repr., Englewood Cliffs, N.J.: Prentice-Hall, 1980), 121.

62. Ibid., 23.

63. Nelkin, "Cultural Perspectives on Blood," 277.

64. Schneider, *American Kinship*, 25.

65. Ibid., 23.

66. Ibid., 24–25.

67. Ibid., 25.

68. Ward Churchill, "The Crucible of American Indian Identity: Native Tradition versus Colonial Imposition in Postconquest North America," *American Indian Culture and Research Journal* 23, no. 1 (1999): 43.

69. Ibid., 47–48.

70. See, for example, J. Philippe Rushton, "The Equalitarian Dogma Revisited," *Intelligence* 19 (1994); Nitzan Mekel-Bobrov et al., "Ongoing Adaptive Evolution of ASPM, a Brain Size Determinant in Homo Sapiens," *Science*, September 9, 2005; and Charlotte Hunt-Grubbe, "The Elementary DNA of Dr. Watson," *London Sunday Times*, October 14, 2007.

71. Jill M. Doerfler, "Fictions and Fractions: Reconciling Citizenship with Cultural Values among the White Earth Anishinaabeg" (PhD diss., University of Minnesota, 2007), 53–60.

72. Ibid., 60.

73. For more discussion on this common analytical mistake, see Wade, *Race, Nature, Culture*.

74. Alexandra Harmon, "Tribal Enrollment Councils: Lessons on Law and Indian Identity," *Western Historical Quarterly* 32 (2001): 177. Harmon refers to critiques made by M. Annette Jaimes, "Federal Indian Identification Policy: A Usurpation of Indigenous Sovereignty in North America," in *The State of Native America: Genocide, Colonization, and Resistance*, ed. M. Annette Jaimes (Boston: South End Press, 1992), 124, 126, and 129; and R. David Edmunds, "Native Americans, New Voices," *American Historical Review* 100 (1995): 733–34.

75. Harmon, "Tribal Enrollment Councils," 179.

76. Ibid., 200.

77. Joseph H. Greenberg, *Language in the Americas* (Stanford: Stanford University Press, 1988).

78. Churchill, "Crucible of American Indian Identity," 41.

79. See Alexandra Harmon, *Indians in the Making: Ethnic Relations and Indian Identities around Puget Sound* (Berkeley: University of California Press, 1998); Harmon, "Tribal Enrollment Councils"; and Dennison, "Constituting an Osage Nation."

80. David Treuer, "How Do You Prove You're an Indian?" *New York Times*, December 20, 2011.

81. Kimberly TallBear, "DNA, Blood, and Racializing the Tribe," *Wicazo Sá Review* 18, no. 1 (2003): 88.

82. For example, see Ward Churchill, *Indians Are Us? Culture and Genocide in Native North America* (Monroe, Maine: Common Courage Press, 1994); Churchill, "Crucible of American Indian Identity"; Winona LaDuke, *All Our Relations: Native Struggles for Land and Life* (Boston: South End Press, 1999); and Michael Yellow Bird, "Decolonizing Tribal Enrollment," in *For Indigenous Eyes Only: A Decolonization Handbook*, ed. Waziyatawin Angela Wilson and Michael Yellow Bird (Santa Fe: School of American Research Press, 2005).

83. For example, see Jaimes, "Federal Indian Identification Policy"; M. Annette Jaimes, "American Racism: The Impact on American-Indian Identity and Survival," in *Race*, ed. Steven Gregory and Roger Sanjek (New Brunswick, N.J.: Rutgers University Press, 1994); J. Kēhaulani Kauanui, *Hawaiian Blood: Colonialism and the Politics of Sovereignty and Indigeneity* (Durham, N.C.: Duke University Press, 2008); Pauline Turner Strong and Barrik Van Winkle, "'Indian Blood': Reflections on the Reckoning and Refiguring of Native North American Identity," *Cultural Anthropology* 11, no. 4 (1996); Gerald Vizenor, *Crossbloods: Bone Courts, Bingo, and Other Reports* (Minneapolis: University of Minnesota Press, 1990); and Yellow Bird, "Decolonizing Tribal Enrollment."

84. See, for example, Leonard A. Carlson, *Indians, Bureaucrats, and the Land: The Dawes Act and the Decline of Indian Farming* (Westport, Conn.: Greenwood, 1980); Churchill, "Crucible of American Indian Identity"; Eva Marie Garroutte,

Real Indians: Identity and the Survival of Native America (Berkeley: University of California Press, 2003); Jaimes, "Federal Indian Identification Policy"; Janet T. McDonnell, *The Dispossession of the American Indian, 1887–1934* (Bloomington: Indiana University Press, 1991); and Theda Perdue, *Nations Remembered: An Oral History of the Five Civilized Tribes, 1865–1907* (Westport, Conn.: Greenwood Press, 1981).

85. See, for example, Joanne Barker, "'Indian-Made' Sovereignty and the Work of Identification" (PhD diss., University of California, Santa Cruz, 2000), 86–88; and Meyer, "American Indian Blood Quantum Requirements," 241.

86. Churchill, "Crucible of American Indian Identity," 49–50.

87. Harmon, *Indians in the Making*, 183–84.

88. Barker, "'Indian-Made' Sovereignty."

89. Kirsty Gover, "Genealogy as Continuity: Explaining the Growing Tribal Preference for Descent Rules in Membership Governance in the United States," *American Indian Law Review* 33, no. 1 (2008): 247.

90. Ibid., 247–48, 286, 292; and Harmon, "Tribal Enrollment Councils," 297. For analyses that emphasize the economic incentives for exclusive tribal-enrollment criteria, also see Meyer, "American Indian Blood Quantum Requirements," 241; and Scott Malcomson, *One Drop of Blood: The American Misadventure of Race* (New York: Farrar, Straus & Giroux, 2000), 115.

91. Lytle and Deloria, *Nations Within*, 246–47.

92. Gover, "Genealogy as Continuity," 257–58.

93. Ibid., 246–47, 251, and 303. I focus on federally recognized tribes because, policy- and resourcewise, there is much at stake. Being a tribal citizen and former planner also conditions my focus on federally recognized tribes. I do, of course, view unrecognized or state-recognized tribes as legitimate targets of inquiry. Dominant conceptions of race shape their identities and capacity for self-government, too. If or when DNA evidence enters federal recognition deliberations, other scholars will certainly tackle how shifting concepts of blood, race, and DNA affect tribes' status.

94. Sturm, *Blood Politics*, 179.

95. Ibid., 179. See also Jodi Byrd, "'Been to the Nation, Lord, but I Couldn't Stay There': Cherokee Freedmen, Internal Colonialism, and the Racialization of Citizenship," in *The Transit of Empire: Indigenous Critiques of Colonialism* (Minneapolis: University of Minnesota Press, 2011); and Melinda Micco, "Blood and Money: The Case of Seminole Freedmen and Seminole Indians in Oklahoma," in *Crossing Waters, Crossing Worlds: The African Diaspora in Indian Country*, ed. Sharon P. Holland and Tiya Miles (Durham, N.C.: Duke University Press, 2006).

96. Churchill, "Crucible of American Indian Identity," 47.

97. Schneider, *American Kinship*.

98. On blood and tribal identity, see, for example, Elizabeth Cook-Lynn, *New Indians, Old Wars* (Urbana: University of Illinois Press, 2007); and Beatrice Medicine, *Learning to Be an Anthropologist*. On the universality of blood invocations, see Meyer, *Thicker Than Water.*

99. Medicine, *Learning to Be an Anthropologist*, 138 and 298.

100. Craig Womack, review of *Anti-Indianism in Modern America: A Voice from Tatekeya's Earth*, by Elizabeth Cook-Lynn, *American Indian Quarterly* 28, nos. 1–2 (2004).

101. Elizabeth Cook-Lynn, *Anti-Indianism in Modern America: A Voice from Tatekeya's Earth* (Urbana: University of Illinois Press, 2001), 79. For a lengthier discussion and analysis of Cook-Lynn's emphasis on tribal nations as "legal entities, rather than merely cultural ones," see Womack, ibid.

102. Cook-Lynn, *New Indians, Old Wars*, 145. Also see, for example, Elizabeth Cook-Lynn, *Why I Can't Read Wallace Stegner and Other Essays: A Tribal Voice* (Madison: University of Wisconsin Press, 1996), 94.

103. Meyer, "American Indian Blood Quantum Requirements"; Philip Deloria, *Playing Indian* (New Haven, Conn.: Yale University Press, 1998); and Shari Huhndorf, *Going Native: Indians in the American Cultural Imagination* (Ithaca, N.Y.: Cornell University Press, 2001).

104. Yellow Bird, "Decolonizing Tribal Enrollment," 180.

105. Meyer, "American Indian Blood Quantum Requirements," 243.

106. TallBear, "Native-American-DNA.coms," 247–48.

107. TallBear, "DNA, Blood, and Racializing the Tribe"; TallBear, ibid.

108. TallBear, "Native-American-DNA.coms," 235–52.

109. Ibid., 236–37.

110. Gover, "Genealogy as Continuity," 250–51.

111. Ibid., 251–52.

112. Gover explains that later in the twentieth century, U.S. tribes were "more likely to include in their constitutions rules prohibiting multiple membership" (ibid., 247–50).

113. Ibid., 252.

114. Kauanui, *Hawaiian Blood*, xi.

115. Ibid., 47–48.

116. Kim TallBear, "The Political Economy of Tribal Citizenship in the U.S.: Lessons for Canadian First Nations?" *Aboriginal Policy Studies* 1, no. 3 (2011).

117. Flyer, DCI America National Tribal Enrollment Conference, October 26–28, 2010, Albuquerque, New Mexico, DCI America, accessed June 11, 2012, http://www.dciamerica.com/pdf/16THATEC.pdf.

2. *The DNA Dot-com*

1. Marks, "What Is Molecular Anthropology?"

2. "Using DNA Testing to Learn about Your Roots," Genetealogy.com, accessed June 12, 2012, http://www.genetealogy.com/. This is a Web site maintained by Megan Smolenyak Smolenyak, professional genealogist and well-known genetic genealogist. However, it does not appear to have been updated since 2005. For a how-to guide for genetic genealogy, see Smolenyak and Turner, *Trace Your Roots with DNA.*

3. See Wolinsky, "Genetic Genealogy Goes Global."

4. See ibid.; Harris, "Genes That Build America"; Simons, "Out of Africa"; Gates, "Promise of Freedom"; Takeaway Media Productions, *Motherland;* Catherine Nash, "Setting Roots in Motion: Genealogy, Geography, and Identity," in *Disputed Territories: Land, Culture, and Identity in Settler Societies*, ed. David Trigger and Gareth Griffiths (Hong Kong: Hong Kong University Press, 2003); and Catherine Nash, "Genetic Kinship," *Cultural Studies of Science Education* 18, no. 1 (2004).

5. See American Society of Human Genetics, "Ancestry Testing Statement," November 13, 2008, accessed February 12, 2013, http://www.ashg.org/pdf/ASHGAncestryTestingStatement_FINAL.pdf. See also Bolnick et al., "Science and Business of Genetic Ancestry Testing."

6. Blaine Bettinger, "How Big Is the Genetic Genealogy Market?" *Genetic Genealogist*, November 6, 2007, accessed February 11, 2013, http://www.thegeneticgenealogist.com/2007/11/06/how-big-is-the-genetic-genealogy-market; and "The Genealogy Market 2009," *Genetic Genealogist*, January 25, 2009, accessed February 11, 2013, http://www.thegeneticgenealogist.com/2009/01/25/the-genealogy-market-2009/.

7. See, for example, Michelle DeArmond, "Man Sues to Gain Admission to Inland Tribe," *Riverside (Calif.) Press-Enterprise*, January 15, 2004; Amy Harmon, "Seeking Ancestry in DNA Ties Uncovered by Tests," *New York Times*, April 12, 2006; Jennifer Kabbany, "Pechanga Denies Disenrolled Family's Appeal," *Californian*, August 18, 2006; and Koerner, "Blood Feud."

8. Margaret Ann Mille, "DNA Print Sells Racial Tests to the Public: The Company Says the Technology Has Forensic and Genealogical Applications," *Sarasota (Fla.) Herald-Tribune*, September 19, 2002.

9. Harmon, "Seeking Ancestry in DNA Ties."

10. Donna Haraway, *Modest_Witness@Second_Millennium.FemaleMan©_Meets_OncoMouse*™ (New York: Routledge, 1997), 129.

11. See, for example, Weiss and Long, "Non-Darwinian Estimation"; Bolnick, "Individual Ancestry Inference"; and Duana Fullwiley, "The Biologistical

Construction of Race: 'Admixture' Technology and the New Genetic Medicine," *Social Studies of Science* 38 (2008).

12. Karl Marx and Friedrich Engels, "The Fetishism of Commodities and the Secrets Thereof," in *The Marx-Engels Reader*, ed. R. C. Tucker (New York: Norton, 1978), 319–21; Haraway, *Modest_Witness*, 142.

13. Haraway, *Modest_Witness*, 142 (emphasis Haraway's).

14. Ibid., 143–44.

15. "DNAPrint Goes Bust," *GenomeWeb Daily News*, GenomeWeb.com, March 3, 2009, accessed June 13, 2012, http://www.genomeweb.com/.

16. "AncestrybyDNA," accessed June 13, 2012, http://www.ancestrybydna.com/.

17. Shriver et al., "Skin Pigmentation," 387.

18. Frank B. Livingstone, "The Duffy Blood Groups, Vivax Malaria, and Malaria Selection in Human Populations: A Review," *Human Biology* 56 (1984); Christophe Tournamille et al., "Disruption of a GATA Motif in the Duffy Gene Promoter Abolishes Erythroid Gene Expression in Duffy-Negative Individuals," *Nature Genetics* 10 (1995); and T. J. Hadley and S. C. Peiper, "From Malaria to Chemokine Receptor: The Emerging Physiologic Role of the Duffy Blood Group Antigen," *Blood* 89 (1997).

19. Martha T. Hamblin and Anna Di Rienzo, "Detection of the Signature of Natural Selection in Humans: Evidence from the Duffy Blood Group Locus," *American Journal of Human Genetics* 66 (2000); and Martha T. Hamblin, Emma E. Thompson, and Anna Di Rienzo, "Complex Signatures of Natural Selection at the Duffy Blood Group Locus," *American Journal of Human Genetics* 70, no. 2 (2002).

20. For a discussion of AIMs, see Deborah A. Bolnick, "'Showing Who They Really Are': Commercial Ventures in Genetic Genealogy" (paper presented at the annual meeting of the American Anthropological Association, Chicago, Ill., 2003); and Fullwiley, "Biologistical Construction of Race."

21. For problems with peer review and transparency in the industry, see also Lee et al., "Illusive Gold Standard."

22. See Esteban J. Parra et al., "Estimating African American Admixture Proportions by Use of Population-Specific Alleles," *American Journal of Human Genetics* 63 (1998); Esteban J. Parra et al., "Ancestral Proportions and Admixture Dynamics in Geographically Defined African Americans Living in South Carolina," *American Journal of Physical Anthropology* 114 (2001); Tony Frudakis et al., "A Classifier for the SNP-Based Inference of Ancestry," *Journal of Forensic Science* 48, no. 4 (2003); and Carrie L. Pfaff et al., "Population Structure in Admixed Populations: Effect of Admixture Dynamics on the Pattern of Linkage Disequilibrium," *American Journal of Human Genetics* 68 (2001).

23. Shriver et al., "Skin Pigmentation," 390.

24. "January 11, 2004," RootsWeb.com GENEALOGY-DNA-L archives, RootsWeb.com, accessed June 13, 2012, http://archiver.rootsweb.ancestry.com/th/read/genealogy-dna/2004-01/1073873510. In personal e-mail correspondence (July 14, 2005), Ann Turner also explained to me her partial analysis of the DNAPrint test as outlined in the January 2004 Genealogy-DNA-L Listserv posts.

25. See Shriver et al., "Skin Pigmentation," 397.

26. See Bolnick, "'Showing Who They Really Are.'"

27. DNAPrint Genomics Web site, quoted in ibid., 4.

28. DNAPrint Genomics Web site, accessed February 12, 2013, http://web.archive.org/web/20060709021118/http://www.ancestrybydna.com/welcome/faq/#q2.

29. "AncestrybyDNA FAQs: What Is Race?" DNAPrint Genomics, accessed February 12, 2013, http://web.archive.org/web/20060709021118/http://www.ancestrybydna.com/welcome/faq/#q1.

30. Koenig, Lee, and Richardson, *Revisiting Race in a Genomic Age;* Reardon, *Race to the Finish.*

31. For example, see Adebowale Adeyemo et al., "A Genome-Wide Association Study of Hypertension and Blood Pressure in African Americans," *PLOS Genetics* 5, no. 7 (2009), doi:10.1371/journal.pgen.1000564; and Rasika A. Mathias et al., "A Combined Genome-Wide Linkage and Association Approach to Find Susceptibility Loci for Platelet Function Phenotypes in European American and African American Families with Coronary Artery Disease," *BMC Medical Genomics* 3, no. 22 (2010), doi:10.1186/1755-8794-3-22.

32. For example, see Esteban G. Burchard et al., "The Importance of Race and Ethnic Background in Biomedical Research and Clinical Practice," *New England Journal of Medicine* 348, no. 12 (2003); and Sally L. Satel, "I Am a Racially Profiling Doctor," *New York Times Magazine*, May 5, 2002.

33. For example, see Pamela Sankar et al., "Genetic Research and Health Disparities," *Journal of the American Medical Association* 291, no. 24 (June 23, 2004).

34. See Nina G. Jablonski, "The Evolution of Human Skin and Skin Color," *Annual Review of Anthropology* 33 (2004): 611–13.

35. Ibid., 612.

36. For additional examples of how research into genetic ancestry and human genetic variation samples selectively between continents (e.g., more rather than less geographically distantly) in order to get greater genetic variation between samples, see Bolnick, "Individual Ancestry Inference," 78–79; Fullwiley, "Biologistical Construction of Race"; and Weiss and Long, "Non-Darwinian Estimation."

37. "AncestrybyDNA," DNAPrint Genomics, accessed September 15, 2005, https://www.ancestrybydna.com/um.asp#um2.

38. "Native American DNA Verification Testing," GeneTree™ DNA Testing Center, accessed March 10, 2006, http://www.genetree.com/product/native-american-test.asp. As of July 2009, this URL reset to http://www.genetree.com/.

39. "James LeVoy Sorenson, World-Renowned Medical Device Inventor, Entrepreneur and Philanthropist, Dies at 86," press release, Sorenson Molecular Genealogy Foundation, accessed June 13, 2012, http://www.smgf.org/press_release.jspx?pr=22.

40. Christine Rosen, "Liberty, Privacy, and DNA Databases," *The New Atlantis*, no. 1 (2003): 46–47.

41. "History," GeneTree.com, accessed June 13, 2012, http://www.genetree.com/history.

42. "Sample Collection Map," Sorenson Molecular Genetics Foundation, accessed June 13, 2012, http://www.smgf.org/maps/collections.jspx.

43. Ibid. My calculations are based on the numbers of samples indicated to date on the Web site when accessed on June 13, 2012.

44. "U.S. POPClock Projection, Component Settings for July 2009," U.S. Census Bureau, accessed July 29, 2009, http://www.census.gov/population/www/popclockus.html.

45. "Sorenson Companies Launch GeneTree, a Unique Genetic-Genealogy Social Networking Web Site," press release, October 23, 2007, GeneTree.com, accessed June 13, 2012, http://www.genetree.com/pressroom.

46. "GeneTree Is Out of Beta: GeneTree's Free Family Site Now Out of Beta with New Interface, Rich Features," Genetree.com blog, accessed June 13, 2012, http://blog.genetree.com/2009/03/genetree-is-out-of-beta.

47. "Scott R. Woodward, Ph.D.," GeneTree.com, accessed June 13, 2012, http://www.genetree.com/.../Scott%20R.%20Woodward%20Biography%20Final.doc.

48. Dan Egan, "BYU Gene Data May Shed Light on Origin of Book of Mormon's Lamanites," *Salt Lake City Tribune*, November 30, 2000. This article indicates Mormon philanthropist Sorenson's interest in the topic of Native American genetic research.

49. Simon G. Southerton, *Losing a Lost Tribe: Native Americans, DNA, and the Mormon Church* (Salt Lake City, Utah: Signature Books, 2004).

50. *Book of Mormon*, 2 Nephi 5:21, for example.

51. *Book of Mormon*, 1 Nephi 13:15, for example.

52. "Native American DNA Verification Testing, GeneTree™ DNA Testing Center," GeneTree.com, accessed June 13, 2012, Internet Archive: Way Back Machine, http://web.archive.org/web/20060615222532/www.genetree.com/product/native-american-test.asp (emphasis GeneTree's).

53. Karen Florin, "The Debate over DNA: Should Pequots Have to Prove Genetic Makeup?" *New London (Conn.) Day*, October 18, 2003.

54. Terry Carmichael, quoted in Adam Tanner, "American Indians Look to DNA Tests to Prove Heritage," *AlertNet*, Thomson Reuters Foundation, March 28, 2005, http://www.trust.org/alertnet/.

55. "Taino and Native American DNA Testing," *CAC Review*, Caribbean Amerindian Centrelink, February 27, 2005, accessed June 13, 2012, http://cacreview.blogspot.com/2005/02/taino-and-native-american-dna-testing.html. FamilyTree DNA also requested that CAC post information about its DNA-testing services.

56. "GeneTree Native American Genetic Ancestry," GeneTree.com, accessed June 13, 2012, Internet Archive: Way Back Machine, http://web.archive.org/web/20060715182305/www.genetree.com/ancestral/nativeAmerican.php.

57. See note 8.

58. "Native American Verification DNA Testing," Genelex, accessed June 13, 2012, Internet Archive: Way Back Machine, http://web.archive.org/web/20100202055931/http://www.healthanddna.com/ancestry-dna-testing/native-american-dna.html.

59. "Native Indian Heritage Testing," Niagen, accessed June 13, 2012, Internet Archive: Way Back Machine, http://web.archive.org/web/20070421184553/http://www.niagen.com/dn_ntvendn_gen_data/dn_native_indian_nav.htm.

60. "Native American Verification DNA Testing."

61. In 2007, Ripan S. Malhi et al. published a paper, "Mitochondrial Haplogroup M Discovered in Prehistoric North Americans," *Journal of Archaeological Science* 34 (2007): 642–48, suggesting that the five-founder model of the settling of the Americas might be insufficient. They published results of an examination of two sets of ancient remains found in China Lake, British Columbia, Canada, that yielded an additional potential "Native American mtDNA haplogroup," labeled M. Haplogroup M has been commonly found in East Asia but "never before . . . in ancient or living indigenous populations in the Americas" (642). However, their results have yet to be replicated by other scientists working on other samples, so there is some doubt as to the durability of their findings.

62. Kim TallBear, field notes, May 2004.

63. "Enrollment Ordinance," ed. Sisseton-Wahpeton Oyate (aka Sisseton-Wahpeton Sioux Tribe) (Sisseton-Wahpeton Oyate, 1982).

64. Steven Whitehead, "The Future of Tribal Enrollment Software" (presentation at the DCI America Tribal Enrollment Conference, New Orleans, La., October 2003. Also see "Genetic Identification Systems, Solutions, Smartcards,"

DNAToday, accessed June 13, 2012, Internet Archive: Way Back Machine, http://web.archive.org/web/20041225035717/www.dnatoday.com/smart_cards.html).

65. Dynamic ID Solutions product list, accessed June 13, 2012, http://www.dynamicidsolutions.com/products.aspx.

66. “Many Tribes Make the Switch to Origins Enrollment Software,” *Industry News*, Tribal Net Online, February 20, 2006, accessed June 13, 2012, http://www.tribalnetonline.com/displaynews.php?newsid=33.

67. “How Would DNA Work for Our Tribe?” DNAToday, accessed June 13, 2012, Internet Archive: Way Back Machine, http://web.archive.org/web/20041227092646/www.dnatoday.com/article_01.html.

68. “Origins Today™,” Dynamic ID Solutions, accessed June 13, 2012, http://www.dynamicidsolutions.com/originstoday.aspx.

69. DCI America home page, accessed June 13, 2012, http://www.dciamerica.com/.

70. DNAToday overhead-projection presentation, DCI America Tribal Enrollment Conference, New Orleans, La., October 2003.

71. TallBear, field notes, DCI America Tribal Enrollment Conference, New Orleans, La., October 28, 2003.

72. Ibid.

73. Ibid.

74. Whitehead, “Future of Tribal Enrollment Software.”

75. DNA Diagnostics Center, PowerPoint presentation at the DCI America National Tribal Enrollment Conference, Albuquerque, N.M., 2010.

76. Gover, “Genealogy as Continuity.”

77. TallBear, field notes, October 28, 2003.

78. Ibid. DNAToday sold two types of paternity tests. One was performed by a “disinterested third party” and could be used to “ascertain legal relationship status.” That was the “legal” paternity test promoted at the DCI America conference. The company also sold, at a lower cost, an at-home test to be used for “information-only situations.”

79. DNA Diagnostics Center, PowerPoint presentation.

80. See “Native American Testing Services,” DNA Diagnostics Center, accessed June 13, 2012, http://www.dnacenter.com/native-american/. I owe a word of thanks to the DDC company representative, Bob Gutendorf, for alerting me to the fact that the term “DNA fingerprint” that I have used, although visually helpful for understanding the technology, is a bit outmoded. It has mostly been replaced in the industry by the term “DNA profile.”

81. “Tribal Testing,” Orchid Cellmark, accessed June 13, 2012, http://www.orchidcellmark.com/tribal.html; and “Aboriginal Testing,” Orchid Cellmark

(Canada), accessed June 13, 2012, http://www.orchidcellmark.ca/site/aboriginal-testing/.

82. In a July 7, 2005, interview with the author, an Orchid Cellmark spokesperson clarified that the company was unique in selling all four types of tests for Native American identity.

83. "Orchid Cellmark Launches New DNA Testing Service to Confirm Native American Tribal Membership," press release, June 17, 2005, Bionity.com, accessed June 13, 2012, http://www.bionity.com/en/news/46940/orchid-cellmark-launches-new-dna-testing-service-to-confirm-native-american-tribal-membership.html.

84. TallBear, interview with Orchid spokesperson, July 7, 2005.

85. Jennifer Clay, interview by author via telephone, July 7, 2005. On July 8, 2005, Clay made a similar comment on *Native America Calling*, a national daily radio show on which she and I were both guests. "DNA Testing for Tribal Enrollment," *Native America Calling*, July 8, 2005; archives available at http://www.nativeamericacalling.com/.

86. Gover, "Genealogy as Continuity."

87. Sam Deloria, "Commentary on Nation-Building: The Future of Indian Nations," *Arizona State Law Journal* 34 (2002); Jessica R. Cattelino, "The Double Bind of American Indian Need-Based Sovereignty," *Cultural Anthropology* 25, no. 2 (2010).

3. Genetic Genealogy Online

1. "Genealogy Is the Fastest Growing Hobby in North America," About.com Genealogy, accessed June 13, 2012, http://genealogy.about.com/library/weekly/aa011502a.htm.

2. Smolenyak and Turner, *Trace Your Roots with DNA*.

3. "Mailing Lists," RootsWeb.com, accessed June 13, 2012, http://lists.rootsweb.com/.

4. Smolenyak and Turner, *Trace Your Roots with DNA;* Nash, "Setting Roots in Motion"; Nash, "Genetic Kinship"; Megan Smolenyak Smolenyak, "'Genetealogy' Survey Results," *Ancestry Daily News*, March 31, 2005.

5. Wolinsky, "Genetic Genealogy Goes Global."; Bolnick et al., "Science and Business of Genetic Ancestry Testing."

6. "How Big Is the Genetic Genealogy Market?" *The Genetic Genealogist*, accessed June 13, 2012, http://www.thegeneticgenealogist.com/2007/11/06/how-big-is-the-genetic-genealogy-market.

7. Smolenyak, "'Genetealogy' Survey Results."

8. Alondra Nelson, "Bio Science: Genetic Ancestry Testing and the Pursuit of African Ancestry," in "Race, Genetics, and Disease: Questions of Evidence, Questions of Consequence," special issue, *Social Studies of Science* (2008); Alondra Nelson, "The Factness of Diaspora," in Koenig, Lee, and Richardson, *Revisiting Race in a Genomic Age;* Harris, "Genes That Build America"; Simons, "Out of Africa"; Gates, "Promise of Freedom."

9. For insight into the popularity of claiming this particular tribal ethnic background, see Circe Sturm, *Becoming Indian: The Struggle over Cherokee Identity in the 21st Century* (Santa Fe, N.M.: School of Advanced Research Press, 2011).

10. Harmon, "Seeking Ancestry"; Koerner, "Blood Feud."

11. Nash, "Setting Roots in Motion"; Nash, "Genetic Kinship"; Nelson, "Bio Science"; Nelson "Factness of Diaspora."

12. Nelson, "Bio Science," 762.

13. Harding, *Sciences from Below,* 31.

14. Blaine Bettinger, *I Have the Results of My Genetic Genealogy Test, Now What?* PDF e-book, *The Genetic Genealogist,* accessed June 14, 2012, http://www.thegeneticgenealogist.com/wp-content/uploads/InterpretingTheResultsofGeneticGenealogyTests.PDF.

15. "NFL Players Xavier Omon and Ogemdi Nwagbuo Confirmed as Half-Brothers," *The Genetic Genealogist,* accessed June 14, 2012, http://www.thegeneticgenealogist.com/2011/09/02/nfl-players-xavier-omon-and-ogemdi-nwagbuo-confirmed-as-half-brothers.

16. Another useful genetic-genealogists' blog is that written by CeCe Moore, *Your Genetic Genealogist,* accessed June 14, 2012, http://www.yourgeneticgenealogist.com/.

17. Bjorn Carey, "First Americans May Have Been European," February 19, 2006, accessed February 19, 2013, http://www.livescience.com/7043-americans-european.html.

18. Bruce Bradley and Dennis Stanford, "The North Atlantic Ice-Edge Corridor: A Possible Palaeolithic Route to the New World," *World Archaeology* 36, no. 4 (2004).

19. The speaker series, which ended in 2005, resulted in the edited volume by Koenig, Lee, and Richardson, *Revisiting Race in a Genomic Age.*

20. Haraway, *Modest_Witness,* 142–44. The way Haraway understands human–nonhuman *social* interactions is another theoretical discussion altogether that I did not get into on-list.

21. James Watson, quoted in Helen Nugent, "Black People 'Less Intelligent,' Scientist Claims," *London Sunday Times,* October 17, 2007.

22. This reaction was also displayed in the on-list responses to a *Science* "Policy Forum" article, "The Science and Business of Genetic Ancestry," by Deborah Bolnick and thirteen other authors, including me, in 2007.

23. Deloria, *Custer Died for Your Sins*, 80–81.

24. Megan Smolenyak Smolenyak, "Don't 'Protect' Us from Our Own Genetic Information," July 20, 2010, accessed February 19, 2013, http://huffingtonpost.com/megan-smolenyak-smolenyak/dont-protect-us-from-our_b_65347.html.

25. Again, see Alondra Nelson's work for data on online communities of black genealogists and genealogists of African descent in both the United Kingdom and the United States: Nelson, "Bio Science"; Nelson, "Factness of Diaspora."

26. Alondra Nelson, e-mail correspondence with author, Fall 2010.

27. For more on the difficulties for African Americans with mixed-race ancestry to assert a mixed-race identity and to have it legitimated by the state, see also Duana Fullwiley, "Race and Genetics: Attempts to Define the Relationship," *Biosocieties* 2 (2007).

28. Stella Ogunwole, *We the People: American Indians and Alaska Natives in the United States*, Census 2000 Special Reports 28 (Washington, D.C.: U.S. Census Bureau, 2006).

29. Sturm, *Becoming Indian*, 58–59.

30. Harmon, "Seeking Ancestry"; Jim Wooten, "Test Suggests 'Black Man' Is Really Not," *ABC Nightline*, December 28, 2003, accessed February 19, 2013, http://abcnews.go.com/Nightline/story?id=129005&page=1.

31. Sturm, *Becoming Indian*, 83 and 87.

32. Ibid., 85.

33. Cheryl I. Harris, "Whiteness as Property," *Harvard Law Review* 106, no. 8 (1993).

34. Jenny Reardon and Kim TallBear, "'Your DNA Is *Our* History': Genomics, Anthropology, and the Construction of Whiteness as Property," *Current Anthropology* 53, no. S5 (2012): S233–S245.

35. Ben-zvi, "Where Did Red Go?" 213.

36. For Morgan's explanation of his theory of "ethnical periods," see Morgan, *Ancient Society*, 3–18. Also see Morgan, *Houses and House-Life*.

37. Ben-zvi, "Where Did Red Go?" 213. Also see Bieder, *Science Encounters the Indian*; and Dippie, *Vanishing American*.

38. Morgan, *Ancient Society*, vii.

39. Morgan, *Houses and House-Life*, xxiv–xxv, 254. In particular, Morgan felt that the study of American aborigines, or "American Indians," should "command as well as deserve the respect of the American people."

40. Ben-zvi, "Where Did Red Go?" 213.

41. Ibid., 203, 208–9.

42. See, for example, Michael Omi and Howard Winant, *Racial Formation in the United States: From the 1960s to the 1990s*, 2nd ed. (New York: Routledge, 1994); Goldberg, *Racist Culture;* and Goldberg, *Racial State.*

43. Ben-zvi, "Where Did Red Go?"; Byrd, "'Been to the Nation'"; Dennison, "Constituting an Osage Nation"; Doerfler, "Fictions and Fractions"; Kauanui, *Hawaiian Blood;* Sturm, *Blood Politics;* Gabrielle Tayac, *IndiVisible: African-Native American Lives in the Americas* (Washington, D.C.: Smithsonian National Museum of the American Indian, 2009); Patrick Wolfe, "Land, Labor, and Difference: Elementary Structures of Race," *American Historical Review* 106, no. 3 (2001).

44. Sturm, *Becoming Indian*, 165–68.

45. Ibid., 57 and 185.

46. Ibid., 55, 61, and 83.

47. Ibid., 54.

48. Ibid., 53.

49. Ibid., 58.

50. Allen, *Origin of Racial Oppression in Anglo-America*, 2; Omi and Winant, *Racial Formation in the United States;* Noel Ignatiev, *How the Irish Became White* (London: Routledge, 1995); David Roediger, *The Wages of Whiteness: Race and the Making of the American Working Class* (London: Verso, 1991).

51. Sturm, *Becoming Indian*, 56–57.

52. Harding, *Sciences from Below;* Susantha Goonatilake, *Toward a Global Science: Mining Civilizational Knowledge* (Bloomington: Indiana University Press, 1998); Banu Subramaniam, "Snow Brown and the Seven Detergents: A Metanarrative on Science and the Scientific Method," in *Women, Science, and Technology: A Reader in Feminist Science Studies*, ed. Mary Wyer et al. (New York: Routledge, 2001); Sturm, ibid.

53. Reardon and TallBear, "'Your DNA Is *Our* History.'"

54. Cheryl Harris, "Theorizing Racial Justice: Reflections on 'Whiteness as Property,' 1993–2013" (keynote address, Fifteenth Annual Dr. Martin Luther King Jr. Commemoration and Luncheon, DePaul University, Chicago, Ill., January 21, 2013).

4. The Genographic Project

1. "Genographic Launches Legacy Fund," May 8, 2006, press release, NationalGeographic.com, accessed June 14, 2012, http://press.nationalgeographic.com/pressroom/pressReleaseFiles.

2. "National Geographic and IBM Launch Landmark Project to Map How Humankind Populated Planet," April 13, 2005, press release, National Geographic.com, accessed June 14, 2012, http://press.nationalgeographic.com/pressroom.

3. "The Genographic Project: Introduction," NationalGeographic.com, accessed June 14, 2012, https://www3.nationalgeographic.com/genographic/index.html.

4. For example, see Deborah Harry and Le`a Malia Kanehe, "Genetic Research: Collecting Blood to Preserve Culture?" *Cultural Survival Quarterly* 29, no. 4 (2006); Elizabeth Pennisi, "Private Partnerships to Trace Human History," *Science*, April 15, 2005, 340; and Henry Greely, "Lessons from the HGDP?" *Science*, June 10, 2005, 1554–55.

5. L. L. Cavalli-Sforza, "Call for a Worldwide Survey of Human Genetic Diversity: A Vanishing Opportunity for the Human Genome Project," *Genomics* 11 (1991): 490–91.

6. *Journey of Man*, directed by Clive Maltby (2003). Spencer Wells is also the author of the related book *The Journey of Man: A Genetic Odyssey* (Princeton, N.J.: Princeton University Press, 2002).

7. "Spencer Wells, Geneticist, Explorer-in-Residence/Emerging Explorer," NationalGeographic.com, accessed June 14, 2012, http://www.nationalgeographic.com/field/explorers/spencer-wells.

8. "Frequently Asked Questions: Question 8," *The Genographic Project*, NationalGeographic.com, accessed June 14, 2012, https://genographic.nationalgeographic.com/genographic/faqs_funding.html#Q8.

9. "National Geographic and IBM Launch Landmark Project."

10. "Frequently Asked Questions: Question 2," *The Genographic Project*, NationalGeographic.com, accessed June 14, 2012, https://www3.nationalgeographic.com/genographic/faqs_privacy.html; Spencer Wells, "Genetic Research: How Much We Have to Learn," *Cultural Survival Quarterly* 29, no. 4 (2006).

11. Reardon, *Race to the Finish*.

12. Rex Dalton, "Tribal Culture versus Genetics," *Nature Genetics* 430 (2004); Rex Dalton, "When Two Tribes Go to War," *Nature Genetics* 430 (2004): 489; Tilousi v. Arizona State University, No. 04-CV-1290 (D.Ariz. 2005); Stephen Hart and Keith Sobraske, *Investigative Report concerning the Medical Genetics Project at Havasupai* (2003), available at the Arizona State University Law Library, Tempe, Ariz.

13. Hart and Sobraske, *Investigative Report*.

14. Tilousi v. Arizona State University, n. 12; Hart and Sobraske, ibid.

15. See Rex Dalton, "Tribe Blasts 'Exploitation' of Blood Samples," *Nature Genetics* 420 (2002); David Wiwchar, "Nuu-chah-nulth Blood Returns to West Coast," *Ha-Shilth-Sa* 31, no. 25 (2005).

16. Spencer Wells, quoted in "The Genographic Project: Introduction," National Geographic.com, accessed October 25, 2005, https://www3.nationalgeographic.com/genographic/index.html. Wells speaks of the "stories" involved in genetics research frequently in his media comments and in his film and accompanying book, *The Journey of Man*. Also see Priscilla Wald, "Blood and Stories: How Genomics Is Rewriting Race, Medicine, and Human History," *Patterns of Prejudice* 40, nos. 4–5 (2006). Wald analyzes population-genetics research, including the Genographic Project, and how it deploys "stories" as technologies in ways that are inseparable from scientific research. Published around the same time as my original Genographic essay, Wald's article analyzes the *Journey of Man* film.

17. Spencer Wells, "We Are All Africans under the Skin," *The Rediff Interview*, November 27, 2002, Rediff.com, accessed June 14, 2012, http://www.rediff.com/news/2002/nov/27inter.htm.

18. "Sample Results," NationalGeographic.com, accessed June 14, 2012, https://www3.nationalgeographic.com/genographic/sample_results.html.

19. Rebecca L. Cann, Mark Stoneking, and Allan C. Wilson, "Mitochondrial DNA and Human Evolution," *Nature*, January 1, 1987, 31–36.

20. Wells, "We Are All Africans."

21. V. Y. Mudimbe, *The Invention of Africa: Gnosis, Philosophy, and the Order of Knowledge* (Bloomington: Indiana University Press, 1988).

22. Spencer Wells, quoted in Simon Collins, "Maori Alarm at Gene Project," *New Zealand Herald*, April 25, 2005.

23. T. Hodgkin, *Nationalism in Colonial Africa* (New York: New York University Press, 1957), quoted in Mudimbe, *Invention of Africa*. In his opening chapter, "Discourse of Power and Knowledge of Otherness," Mudimbe references Hodgkin's classic treatment of the politics of late colonial Africa, particularly the Rousseauian image of the continent. Mudimbe notes the coexistence of seemingly contradictory myths of African otherness.

24. Peter Dwyer, quoted in Robert N. Proctor, "Three Roots of Human Recency: Molecular Anthropology, the Refigured Acheulean, and the UNESCO Response to Auschwitz," *Current Anthropology* 44, no. 2 (2003).

25. Joanne Barker, "The Human Genome Diversity Project: 'Peoples,' 'Populations,' and the Cultural Politics of Identification," *Cultural Studies of Science Education* 18, no. 1 (2004); Lisa Gannett, "Making Populations: Bounding Genes in Space and in Time," *Philosophy of Science* 7 (2003); Lisa Gannett, "Racism and Human Genome Diversity Research: The Ethical Limits of 'Populational Thinking,'" *Philosophy of Science* 68, no. 3, supplement (2001); Jennifer Reardon, "The Human Genome Diversity Project: A Case Study in Coproduction," *Social Studies of Science* 31, no. 3 (2001); Reardon, *Race to the Finish*.

26. Wald, "Blood and Stories," 319.

27. Wells, "We Are All Africans."

28. See Gannett, "Making Populations"; Gannett, "Racism and Human Genome Diversity Research"; Gould, "American Polygeny and Craniometry"; Marks, *Human Biodiversity;* Reardon, "Human Genome Diversity Project"; Reardon, *Race to the Finish;* Stepan, *The Idea of Race in Science;* and Stocking, *Race, Culture, and Evolution.*

29. "National Geographic, Genographic Project Director Spencer Wells, IBM Lead Scientists Ajay Royyuru Answer Questions about the Project," press release, National Geographic.com, accessed June 14, 2012, http://press.nationalgeographic.com/pressroom/pressReleaseFiles/1113322781773/1113322781790/Geno_Q&A_Final%204.26.06.pdf. Steve Olson, "The Genetic Archaeology of Race," *Atlantic,* April 2001, 80, noted the same naive expectation by Wells's mentor L. Luca Cavalli-Sforza that population genetics can end racism.

30. Wells, *Journey of Man* (film).

31. Lewis Henry Morgan, *Ancient Society; or, Researches in the Line of Human Progress from Savagery through Barbarism to Civilization* (1877; repr., Chicago: Kerr, 1909), vii.

32. Human Genome Diversity Project, *Human Genome Diversity Workshop 1* (Stanford, Calif.: Stanford University Press, 1992); Human Genome Diversity Project, *Human Genome Diversity Workshop 2* (State College, Pa.: Pennsylvania State University, 1992); L. Luca Cavalli-Sforza et al., "Call for a Worldwide Survey of Human Genetic Diversity: A Vanishing Opportunity for the Human Genome Project," *Genomics* 11, no. 2 (1991): 490; Reardon, *Race to the Finish,* 105.

33. See chapter 1 for a fuller discussion of Boas's influence on the "culture" concept in anthropology and how it confounded notions of biological race in twentieth-century scholarship.

34. See Waziyatawin Angela Wilson, with translations from the Dakota text by Wahpetunwin Carolynn Schommer, *Remember This! Dakota Decolonization and the Eli Taylor Narratives* (Lincoln: University of Nebraska Press, 2005). Non-Dakota thinkers have also treated the 1862 war at length, providing valuable insights, particularly into U.S. Indian policy that led to the war. For example, see Gary Clayton Anderson, *Little Crow, Spokesman for the Sioux* (St. Paul: Minnesota Historical Society Press, 1986); and Roy Willard Meyer, *History of the Santee Sioux: United States Indian Policy on Trial* (Lincoln: University of Nebraska Press, 1967). But when I say that 1862 figures prominently in how we understand what it is to be Dakota today, I do not want to mislead the reader by citing non-Dakota thinkers. We do not acquire our understanding of who we are in relation to 1862 by reading the necessarily narrow works of non-Dakota historians. The various Dakota communities throughout the upper Midwest and

parts of Canada since 1862 have lived a history largely untreated in the scholarly literature. That is why the book by Wilson and Schommer—two Dakota thinkers—is so important. Throughout Dakota country, we refer daily to 1862, and Wilson and Schommer foreground Dakota narratives of that history in a written text. The many other citations that should be here—for example, narratives handed down from my great-grandfather (his great-grandfather was Te-oyate-duta, a reluctant leader of the Dakota effort in 1862) to my mother and to many other family members—are largely undocumented.

35. See "National Geographic and IBM Launch Landmark Project."

36. Debra Harry and Frank C. Dukepoo, *Indians, Genes, and Genetics: What Indians Should Know about the New Biotechnology* (Nixon, Nev.: Indigenous Peoples Coalition against Biopiracy, 1998), 24.

37. David Hurst Thomas, *Skull Wars: Kennewick Man, Archaeology, and the Battle for Native American Identity* (New York: Basic Books, 2000).

38. Native American Graves Protection and Repatriation Act (NAGPRA), Pub. L. No. 101–601, 25 U.S.C. 3001 et seq., 104 Stat. 3048 (1990), § 7a(4).

39. Bonnichsen et al. v. United States et al., 2004-02-04, no. 02-35994 (9th Cir. 2004), 3.

40. Francis P. McManamon, "The Initial Scientific Examination, Description, and Analysis of the Kennewick Man Human Remains," in *Report on the Non-destructive Examination, Description, and Analysis of the Human Remains from Columbia Park, Kennewick, Washington* (Washington, D.C.: National Park Service, 1999).

41. Aviva L. Brandt, "Suit over Kennewick Man Revived," *Seattle Post-Intelligencer*, October 26, 2000; James Chatters, "Encounter with an Ancestor," *Anthropology Newsletter* 38, no. 1 (1997): 9–10; Steve Coll, "The Body in Question," *Washington Post*, June 3, 2001; Mike Lee, "Recasting the Past: Day Three; No Turning Back on the Kennewick Man," Hypography.com, http://www.hypography.com/; Mike Lee, "Politics of the Past," *Kennewick (Wash.) Tri-City Herald*, December 26, 1999.

42. Federica A. Kaestle, "Chapter 2: Report on DNA Analysis of the Remains of 'Kennewick Man' from Columbia Park, Washington," in *Report on the DNA Testing Results*. Also see David G. Smith et al., "Chapter 4: Report on DNA Analysis of the Remains of 'Kennewick Man' from Columbia Park, Washington," in the same publication.

43. Gover, "Genealogy as Continuity."

44. Bonnichsen et al. v. United States et al.

45. Smith et al., "Chapter 4: Report on the DNA Testing Results."

46. "Frequently Asked Questions: Question 7," NationalGeographic.com, accessed June 17, 2012, https://www3.nationalgeographic.com/genographic/faqs

_about.html; and "Back from Chad!," NationalGeographic.com, accessed June 17, 2012, https://www3.nationalgeographic.com/genographic/resources.html.

47. "Geno 2.0—Genographic Project Participation and DNA Ancestry Kit," National Geographic.com, accessed February 19, 2013, http://shop.nationalgeographic.com/ngs/browse/productDetail.jsp?productId=2001246.

48. "The Genographic Legacy Fund: Building Capacity, Raising Awareness," NationalGeographic.com, accessed June 17, 2012, https://www3.nationalgeographic.com/genographic/legacy_fund.html.

49. "IPCB Comments regarding the Genographic Project's Request for Proposals for the Legacy Fund," February 16, 2006, Indigenous Peoples Council on Biocolonialism, accessed June 17, 2012, http://www.ipcb.org/issues/human_genetics/htmls/legacyfund_rfp.html. The IPCB's e-mail notice was forwarded to me by multiple indigenous contacts globally during 2006 and early 2007.

50. "Genographic Legacy Fund."

51. Ibid.

52. "Indigenous Representatives Talk about Their Migratory Histories" (video), NationalGeographic.com, accessed October 25, 2005, https://www3.nationalgeographic.com/genographic/index.html. When I attempted to access the video on December 10, 2005, the link to the video clip was no longer available. Repeated requests for information on how to obtain a copy of the video have gone unfulfilled.

53. Reardon, *Race to the Finish*, especially 115–22.

54. Faye Flam, "Sweeping DNA Project Aims to Chart Human History," *Wilkes-Barre (Pa.) Times Leader*, September 5, 2005.

55. "The Genographic Project: A Landmark Study of the Human Journey," Seaconke Wampanoag Tribe, accessed June 17, 2012, http://kalel1461.tripod.com/id17.html.

56. Ibid.

57. James Clifford, "Identity in Mashpee," in *The Predicament of Culture: Twentieth-Century Ethnography, Literature, and Art* (Cambridge, Mass.: Harvard University Press, 1988), 277–346.

58. Ibid., 285.

59. "Favorite Tribal Photos," Seaconke Wampanoag Tribe, accessed June 17, 2012, http://kalel40.tripod.com/id2.html.

60. The tribal history referred to was published on the tribe's Web site, Seaconkewampanoagtribe.net (last visited July 17, 2007), which was off-line when checked on July 9, 2009. The current Web site has a different history published, which largely precedes white settlement in the area. See "History: Seaconke Wampanoag Tribe," accessed June 17, 2012, http://kalel1461.tripod.com/id23.html.

61. Russ Olivo, "Tribe: Cumberland, Woonsocket Are Ours," *Woonsocket (R.I.) Call*, January 26, 2003.

62. Clifford, "Identity in Mashpee," 289.

63. Anne Merline McCulloch and David E. Wilkins, "Constructing Nations within States: The Quest for Federal Recognition by the Catawba and Lumbee Tribes," *American Indian Quarterly* 19, no. 3 (1995): 361–88.

64. Jasanoff, *States of Knowledge*, 2.

65. Ibid.

66. Zhadanov et al., "Genetic Heritage and Native Identity."

67. Ibid., 582.

68. Ibid., 586.

69. See Fortmann, "Gendered Knowledge"; and Chidari et al., "Use of Indigenous Trees in Mhondoro District."

Conclusion

1. Meyer, *Thicker Than Water*.

2. Indian and Native American Employment and Training Coalition, *Counting Indians in the 2000 Census: Impact of the Multiple Race Response Question*, special report (Washington, D.C.: Indian and Native American Employment and Training Coalition, 2004); Ogunwole, *We the People*.

3. Reardon and TallBear, "'Your DNA Is *Our* History.'"

4. Patricia Mariella et al., "Tribally-Driven Participatory Research: State of the Practice and Potential Strategies for the Future," *Journal of Health Disparities Research and Practice* 3, no. 2 (2009); "The Mataatua Declaration on Cultural and Intellectual Property Rights of Indigenous Peoples, June 18, 1993," in *Pacific Genes and Life Patents: Pacific Indigenous Experiences and Analysis of the Commodification and Ownership of Life*, ed. Aroha Te Pareake Mead and Steven Ratuva (Wellington, N.Z.: Call of the Earth [Llamado de la Tierra] and the United Nations University Institute of Advanced Studies, 2007), as well as the rest of Mead and Ratuva's volume; Aroha Mead, "Genealogy, Sacredness, and the Commodities Market," *Cultural Survival Quarterly* 20, no. 2 (Summer 1996): 46–51. The Mataatua Declaration was passed by a plenary of delegates from Ainu (Japan), Australia, the Cook Islands, Fiji, India, Panama, Peru, the Philippines, Surinam, the United States, and Aotearoa (New Zealand).

5. Goering, Holland, and Fryer-Edwards, "Transforming Genetic Research Practices"; Reardon and TallBear, "'Your DNA Is *Our* History'"; "Genomics, Governance, and Indigenous Peoples," workshop, accessed June 17, 2012, http://cnr.berkeley.edu/tallbear/workshop/index.html.

6. Tsosie, "Cultural Challenges to Biotechnology," 396–97, 405–9.

7. Ibid., 397–98.

8. Ibid., 401.

9. Moore v. Regents of the University of California, 271 Cal. Rptr. 146, 401 (1991).

10. Tsosie, "Cultural Challenges to Biotechnology," 397, 398.

11. Ibid., 408–9.

12. *CIHR Guidelines for Health Research Involving Aboriginal People*, Canadian Institutes of Health Research, accessed June 18, 2012, http://www.cihr-irsc.gc.ca/e/29134.html.

13. Tsosie, "Cultural Challenges to Biotechnology," 408–9.

14. "Fiscal Year 2009 Tribal Consultation Report," Centers for Disease Control and Prevention, accessed January 3, 2012, http://www.cdc.gov/omhd/reports/2009/CDCTBCR2009.pdf.

15. Terry Powell, "Genomics, Tribes, and Indigenous Peoples" workshop, 2008. This workshop, funded by the National Science Foundation and hosted by the Arizona State University Law School and its American Indian Policy Institute, November 6–7, 2008, consisted of conversations and strategizing about indigenous genomics and representation, sovereignty, and property. Participants, in addition to me, included legal scholars and practitioners Philip "Sam" Deloria (American Indian College Fund), Nadja Kanellopoulou (Oxford University), Pilar Ossorio (University of Wisconsin), Brett Lee Shelton and Rebecca Tsosie (ASU Law); science studies scholars Paul Oldham and Brian Wynne (Lancaster University, U.K.); geneticists Laura Arbour (University of British Columbia) and Nanibaa' Garrison (Stanford University); Native American IRB expert Terry Powell (Alaska Area Indian Health Service); and Jenny Reardon (UC Santa Cruz). The workshop was the first in a series of meetings and related projects that will explore opportunities for expanding indigenous governance of genomic research.

16. "Fiscal Year 2009 Tribal Consultation Report."

17. Laura Arbour and Doris Cook, "DNA on Loan: Issues to Consider When Carrying Out Genetic Research with Aboriginal Families and Communities," *Community Health Genetics* 9 (2006): 155. Also see Jennifer Couzin-Frankel, "Researchers to Return Blood Samples to Yanomamö," *Science*, June 4, 2010, 1218.

18. Reardon and TallBear, "'Your DNA Is *Our* History.'"

19. See note 15 for names of collaborating scholars.

20. For such models, see American Indian Law Center, *Model Tribal Research Code: With Materials for Tribal Regulation for Research and Checklist for Indian Health Boards*, 3rd ed. (Albuquerque, N.M.: American Indian Law Center, 1999); Indigenous Peoples Council on Biocolonialism, *Indigenous Research Protection Act* (Wadsworth, Nev.: Indigenous Peoples Council on Biocolonialism, 2000).

21. "Genomics, Governance, and Indigenous Peoples."

22. "Science and Justice Working Group," University of California, Santa Cruz, accessed June 18, 2012, http://scijust.ucsc.edu/.

23. "Science and Justice Training Program," University of California, Santa Cruz, accessed June 18, 2012, http://research.pbsci.ucsc.edu/scienceandjustice/blog/ethics-and-justice-in-science-and-engineering-training-program.

24. *Tri-Council Policy Statement: Ethical Conduct for Research Involving Humans*, 2nd ed., Government of Canada Panel on Research Ethics, accessed June 18, 2012, http://www.pre.ethics.gc.ca/eng/policy-politique/initiatives/tcps2-eptc2/Default/.

25. Laura Arbour, personal conversation with author, June 1, 2012.

26. Paul C. Webster, "Canada Curbs Aboriginal Health Leadership," *Lancet*, June 9, 2012, 2137.

27. *CIHR Guidelines.*

28. Samuel W. Anderson, "Sacred Grounds: A Community of 'the Last Incas' Bristles at National Geographic's Attempt to Collect Their DNA," *Gene Watch* 24, no. 2 (2011); Lynn Hermann, "Indigenous Tribe from Peru Stops Nat'l Geographic's DNA Sampling," *Digital Journal* (2011); Antonio Regalado, "Indigenous Peruvian Tribe Blocks DNA Sampling by National Geographic," *ScienceInsider*, May 6, 2011, accessed February 19, 2013, http://news.sciencemag.org/scienceinsider/2011/05/indigenous-peruvian-tribe-blocks.html.

29. Asociación ANDES, "ANDES Communiqué, May 2011: Genographic Project Hunts the Last Incas," KimTallBear.com, accessed June 18, 2012, http://www.kimtallbear.com/uploads/5/3/1/5/5315525/andes_communique_genographic_project_hunts_the_last_incas.pdf. The original link, http://64.22.85.140/%7Ecommuniq/pdf/ANDES_Communique_Genographic_Project_Hunts_the_Last_Incas.pdf, no longer yields the document.

30. I first blogged about the Asociación ANDES communiqué on May 12, 2011. See "Genographic Back in the News: Badly Organized Genetic Sampling of Indigenes in Peru," accessed June 18, 2012, http://www.kimtallbear.com/1/post/2011/05/genographic-back-in-the-news-badly-organized-genetic-sampling-of-indigenes-in-peru.html. This final section of the chapter incorporates that blog entry.

31. See Regalado, "Indigenous Peruvian Tribe."

32. "ANDES Communiqué," 3–5.

33. Ibid., 4.

34. Ibid., 3.

35. For a link to the letter, see Regalado, "Indigenous Peruvian Tribe."

36. "ANDES Communiqué," 8–9.

37. Rasmus Nielsen, "Do the Genes Belong to the Tribal Council?" September 30, 2011, *Evolutionary Genomics Blog*, accessed June 18, 2012, http://cteg.berkeley.edu/~nielsen/blog/.

38. Dennis O'Rourke, "Anthropological Genetics in the Genomic Era: A Look Back and Ahead," *American Anthropologist* 105, no. 1 (2003): 104.

39. "ANDES Communiqué," 3–4.

40. Ibid., 2–3.

41. "Ethical Framework," Genographic Project, accessed June 18, 2012, https://genographic.nationalgeographic.com/staticfiles/genographic/StaticFiles/AboutGenographic/Introduction/Genographic-Project-Ethics-Overview.pdf.

42. Laura Arbour, personal conversation with author, June 1, 2012.

43. Dalton, "Tribe Blasts 'Exploitation' of Blood Samples."

44. Jana Bommersbach, "Arizona's Broken Arrow," *Phoenix Magazine*, November 2008; Amy Harmon, "Indian Tribe Wins Fight to Limit Research of Its DNA," *New York Times*, April 21, 2010.

45. Harmon, "DNA Gatherers Hit Snag."

46. Indigenous Peoples Council on Biocolonialism, *Indigenous People, Genes, and Genetics: What Indigenous Peoples Should Know about Biocolonialism; A Primer and Resource Guide*, http://www.ipcb.org/publications/primers/htmls/ipgg.html.

47. Kimberly TallBear, "Narratives of Race and Indigeneity in the Genographic Project," *Journal of Law, Medicine, and Ethics* 35, no. 3 (2007).

48. John Trudell, "We Are the Shapes of the Earth," *DNA: Descendant Now Ancestor*, ASITIS Productions / Effective Records, 2001, compact disc.

49. Reardon and TallBear, "'Your DNA Is *Our* History.'"

50. Jenny Reardon, "The Democratic, Anti-racist Genome? Technoscience at the Limits of Liberalism," *Science as Culture* 21, no. 1 (2012).

INDEX